CHROMATOGRAPHIA

An International Journal for Rapid Communication in Chromatography, Electrophoresis, and Associated Techniques

Abstracted in Anal. Abstr., ASCA. Biodet. Abstr., Biol. Abstr., Cadscan, Chem. Abstr., Chem. Cit. Ind., C.I.S. Abstr., Current Contents, Deep Sea Res. & Oceanogr. Abstr., Diary Sci. Abstr., Excep. Med., Food Sci. & Tech. Abstr., GeoRef., INIS Atormind. Ind. Sci. Rev., Ind. Vet., Lead Abstr., Mass Spectr. Bull., Nat. Sci. Cit. Ind., Rev. Med. & Vet. Mycol., Sci. Cit. Ind., Sel. Water Res. Abstr., Sugar Ind. Abstr., Vet. Bull., VITIS, Weed Abstr., W.R.C. Inf., Zinc Scan

Volume 51, Supplement, 2000

Editorial Office	M. Schaub, Manager	W. Schwarz, In-House Editor	H. Weinheimer, Publisher
	Vieweg Publishing P O Box 1546 65173 Wiesbaden, Germany		
	Tel. +49 (0)611 7878 380, 381 Fax +49 (0)611 7878 439		

vieweg

CHROMATOGRAPHIA

An International Journal for Rapid Communication in Chromatography, Electrophoresis, and Associated Techniques

Contents Supplement, Volume 51, 2000

Publisher

Chromatographia is published by Friedr. Vieweg & Sohn Verlagsgesellschaft mbH, P.O. Box 1546, D-65173 Wiesbaden, Federal Republic of Germany, Tel. +49 (0)611 7878 380(-381); Telefax +49 (0)611 7878 439

Editorial office e-mail: CHROMATOGRAPHIA@bertelsmann.de

For more information regarding Vieweg's program for books and journals see our homepage: http://www.vieweg.de

Advertising Representatives

Inquiries concerning advertising should be addressed to the publisher's address above; Tel. +49 (0)611 7878 153, Fax –430.

Inquiries in USA: Trade Media International, 424 Madison Avenue, New York, NY 10017, USA; Tel. (212) 421–1229.

Inquiries in the UK: Elsevier Science Ltd., The Boulevard, Langford Lane, Kidlington, Oxford, OX5 1GB, UK.

Distributors

Friedr. Vieweg & Sohn, P.O. Box 1546, D-65173 Wiesbaden, Germany; Tel. +49 (0)611 7878 324; Telefax +49 (0)611 7878 423.

Elsevier Science Ltd., The Boulevard, Langford Lane, Kidlington, Oxford, OX5 1GB, UK.

Distributions in the USA

Chromatographia (USPS No. 374 810) is distributed by German Language Publications, Inc., 153 South Dean Street, Englewood, NJ 07631. Second class postage is paid at Englewood, NJ 07631.

Postmaster: send address changes to Chromatographia, German Language Publications, Inc., 153 South Dean Street, Englewood, NJ 07631.

Subscriptions

Chromatographia is published monthly. Up to three volumes may be published per year.

Vols. 51, 52, Supplements (2000) DM 2.085,–   US $ 1.036.00
(approx. 2150 pp.)   öS 15.221,–   sFr   1.855,–
Single copy   DM 198,–   US $ 98.00
                                   öS 1.445,–   sFr   76,–

For individual subscribers who will certify that Chromatographia is for their personal use only (to be ordered directly from the publisher):

Vols. 51, 52, Supplements (2000) DM 1.182,–   US $ 588.00
                                   öS 8.629,–   sFr   1.051,–

All prices include postage. Subscriptions are renewed automatically for one year unless notice to terminate the subscription is given three months before the end of the current year.

Submission of Papers

One original and two copies of manuscripts should be sent to Chromatographia, Editorial Office at the same address as the publisher, above. For papers intended for review-type articles, an outline of the proposed article should first be forwarded to the Editorial Office Manager for preliminary discussion, prior to preparation. For general information on the rules concerning style and format of manuscripts please refer to ,,Instructions to Authors" in every issue.

Chromatographia was founded by R. E. Kaiser in 1968.

Vieweg is a company in the specialist publishing group BertelsmannSpringer

© Friedr. Vieweg & Sohn Verlagsgesellschaft mbH, Braunschweig/Wiesbaden, 2000

Softcover reprint of the hardcover 1st edition 2000

ISBN-13: 978-3-322-83135-4   e-ISBN-13: 978-3-322-83133-0
DOI: 10.1007/978-3-322-83133-0

ISSN 0009-5893

Chromatogram on front page: Gas chromatographic separation of gasoline hydro carbons (selected section of chromatogram) with glass capillary column.

Preface

In the almost 20 years since its first appearance, Capillary Electrophoresis (CE) has become a standard analytical separation technique in many laboratories. CE is now used routinely in applications ranging from inorganic ion determinations to genetic analysis. After the first development studies in R&D groups in the first 10 years, and the breakthrough of commercial instruments in the second decade, CE has now become a member of the establishment of analytical techniques. This seemed a good moment to summarize the instrumental possibilities of the technique, its accomplishments and weak points. The result of this thought is in front of you.

The reader will not find much chemistry in this monograph. Subjects such as micellar phases, chiral discrimination, or Ogston sieving regimes will be discussed in other publications in this series. As the title already indicates, I have focused on the instrumentation and system aspects. Within the system I include the background electrolyte, an often underestimated factor in electrophoretic separations.

The various paragraphs and chapters differ somewhat in character. When the subject of a paragraph is something well established and the topics discussed can be regarded as common knowledge, the text approaches that of a student textbook; as an extra, I have included some (hopefully) good advice for those who are mainly interested in the practical application of CE. Paragraphs on subjects that are still new and under investigation resemble more a literature review. I have indicated the direction of research and developments in these areas, without pretending to know where it will end.

Some instrumental aspects have not received the attention they deserve. The most striking examples are CE-MS coupling and "CE-on-a-chip". This is for two reasons. First, these subjects are still hot topics in research, and any review will have only a limited lifetime. Secondly (and more importantly) I have not enough knowledge and experience in these areas to write more than a short summary of what is going on.

Writing a monograph like this is a lot of work, even more than I anticipated. Luckily, I could fall back on the theoretical and practical experience with CE obtained in our laboratory over the past years. Therefore, I am indepted to all who contributed to this experience. In the first place my (former) colleagues Hans Poppe and Johan Kraak should be mentioned. They started the CE work in Amsterdam and taught me a lesson or two about it later. Together we formed maybe not the most famous but certainly the best-spirited CE research group in the world. Thanks are also due to the hard-working lab members Sytske Heemstra and Wim Ozinga; they did a lot of the experimental work supporting the theory in this manuscript. As you will see later, I borrowed many ideas and experimental data from the work of (Ph.D.) students and guests of the Laboratory for Analytical Chemistry in Amsterdam. For this I want to thank in particular: Amit Asthana, Dev Bose, Michel van Bruijnsvoort, Gerard Bruin, Alejandro Cifuentes, Le Thi Huyen Duong, Abilasha Durgbanshi, Anupama Gaur, Xinjian Huang, Yüksel Sahin, Sunil Sanghi, Reza Maleki Seifar, Remco Stol, Anna Tüdös, Xiaoma Xu and Ruohua Zhu, and all others who have been involved in CE research here.

Finally, the editor of the series of which this monograph is one, Kevin Altria, deserves compliments for his patience and his friendly way of putting me under pressure.

Amsterdam, December 1999

Wim Th. Kok

Laboratory for Analytical Chemistry

Department of Chemical Engineering

University of Amsterdam

1 The Short History of CE

Capillary electrophoresis (CE) is a relatively young technique with a firm historical foundation. The first appearance of CE, in a configuration essentially the same as we know it today, can be found in 1981 with the publication of an article by Jorgenson and Lukacs, working at the University of North Carolina (USA), in Analytical Chemistry [1]. They showed the separation of dansyl and fluorescamine derivatized amino acids, peptides and amines in a Pyrex glass capillary with an internal diameter of 75 μm, using a 30 kV voltage source to drive the electrophoretic separation.

It was of course already known for a long time that an electric field over a solution can be used to separate charged particles. In the 19th century, physical chemists had applied electrophoretic methods to separate or fractionate colloids. The theoretical basis for the description of the electrophoresis of ions was laid by F. Kohlrausch in 1897. A copy of the first page of his original article in Annalen der Physik und Chemie [2], in which he formulated the "beharrliche Funktion" or regulating function for electrophoresis, is shown as Figure 1.1. These Kohlrausch' functions are still essential in CE to understand phenomena such as overloading, sample stacking and indirect detection. Tiselius developed his moving boundary electrophoresis in the period of 1930–1940 [3]. A further extension of the theory of moving and stagnant zones in electrophoresis, including multiple ion systems, was given by Dole [4] and by Longworth [5].

In these early days electrophoresis was used mostly as a tool to study the behaviour of electrolyte solutions and to characterise charged particles such as colloids and proteins. However, the possibilities of electrophoretic methods for analytical and preparative separations were already understood; in the words of Tiselius [6]: "It appeared very tempting to utilize the efficient separation, which can be observed optically in the moving boundary methods, in order to isolate the components for further investigation of their chemical and biochemical properties (...). The overlapping "boundary separation" had thus to be substituted by a "zone separation" in which each component was allowed to form a zone separated from the others by empty regions."

The main bottleneck was the convection in the solution due to density differences between zones and the "empty regions". Application of electrophoresis as a routine analytical separation method only became possible with the introduction of a stabilising medium for the separation electrolyte, preventing the detrimental effects of convection in the solution. In first instance (around 1950) paper strips were used as the separation medium. In later decennia agar and polyacrylamide gels were introduced. Nowadays, in clinical laboratories (slab) gel electrophoresis is still of utmost importance. Despite the success of slab gel electrophoresis, some disadvantages of the technique were also recognised.

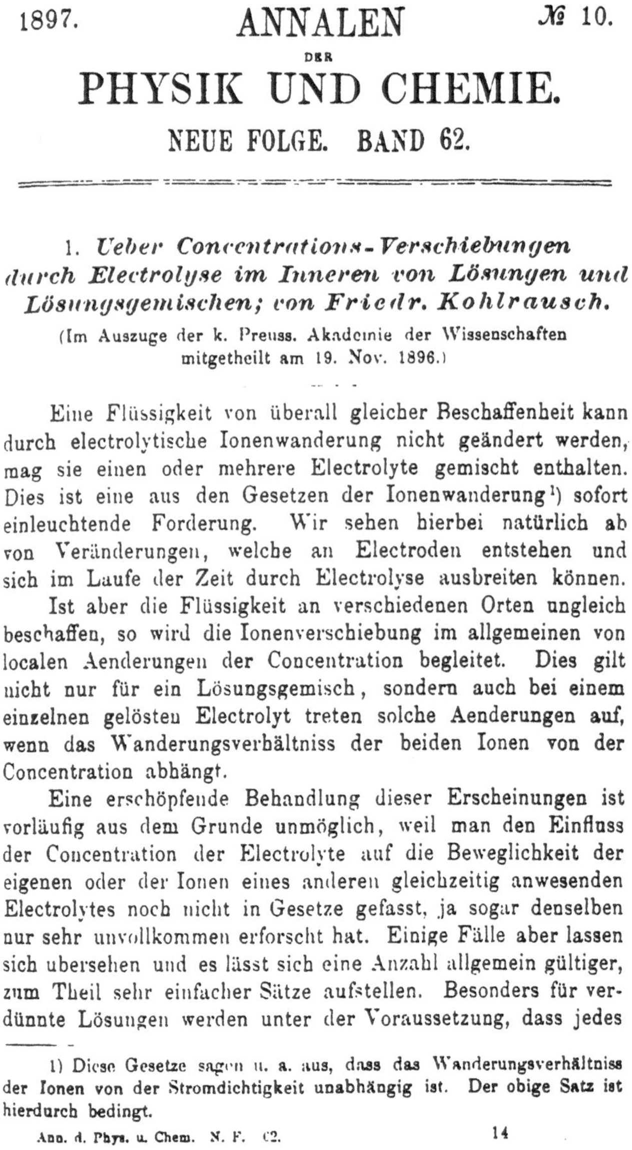

1897.   ANNALEN
DER
PHYSIK UND CHEMIE.
NEUE FOLGE. BAND 62.

№ 10.

1. *Ueber Concentrations-Verschiebungen durch Electrolyse im Inneren von Lösungen und Lösungsgemischen; von Friedr. Kohlrausch.*
(Im Auszuge der k. Preuss. Akademie der Wissenschaften mitgetheilt am 19. Nov. 1896.)

Eine Flüssigkeit von überall gleicher Beschaffenheit kann durch electrolytische Ionenwanderung nicht geändert werden, mag sie einen oder mehrere Electrolyte gemischt enthalten. Dies ist eine aus den Gesetzen der Ionenwanderung[1] sofort einleuchtende Forderung. Wir sehen hierbei natürlich ab von Veränderungen, welche an Electroden entstehen und sich im Laufe der Zeit durch Electrolyse ausbreiten können.

Ist aber die Flüssigkeit an verschiedenen Orten ungleich beschaffen, so wird die Ionenverschiebung im allgemeinen von localen Aenderungen der Concentration begleitet. Dies gilt nicht nur für ein Lösungsgemisch, sondern auch bei einem einzelnen gelösten Electrolyt treten solche Aenderungen auf, wenn das Wanderungsverhältniss der beiden Ionen von der Concentration abhängt.

Eine erschöpfende Behandlung dieser Erscheinungen ist vorläufig aus dem Grunde unmöglich, weil man den Einfluss der Concentration der Electrolyte auf die Beweglichkeit der eigenen oder der Ionen eines anderen gleichzeitig anwesenden Electrolytes noch nicht in Gesetze gefasst, ja sogar denselben nur sehr unvollkommen erforscht hat. Einige Fälle aber lassen sich übersehen und es lässt sich eine Anzahl allgemein gültiger, zum Theil sehr einfacher Sätze aufstellen. Besonders für verdünnte Lösungen werden unter der Voraussetzung, dass jedes

1) Diese Gesetze sagen u. a. aus, dass das Wanderungsverhältniss der Ionen von der Stromdichtigkeit unabhängig ist. Der obige Satz ist hierdurch bedingt.
Ann. d. Phys. u. Chem. N. F. 62.   14

Figure 1.1
Copy of the first page of the publication by Kohlrausch [2] in 1897 on the regulating function in electrophoresis.

Generally, analysis times tended to be long. The separation voltage had to be kept relatively low to avoid overheating of the gels. Also, quantitative interpretation of the results was problematic because the separated compounds had to be detected *in situ* on the gel.

Before the publication of the work of Jorgenson and Lukacs, studies had been carried out in different European laboratories on electrophoresis in free solution, that paved the way for modern CE. Hjerten (University of Uppsala, Sweden) in 1967 [7] showed the importance of the suppression of convective effects, induced by the generated heat during the electrophoresis, for the separation efficiency. Since he used relatively wide, glass separation tubes, he proposed a system in which the tube was rotated around its axis. Virtanen in 1974 [8] showed the advantages of using narrow capillaries for the separation efficiency. In the seventies, pioneering work

0009-5893/00   S-6-03 $ 03.00/0   © 2000 Friedr. Vieweg & Sohn Verlagsgesellschaft mbH

had been performed by the group of Everaerts at the Technical University of Eindhoven (Netherlands) on isotachophoresis [9]. In 1979 two articles by this group were published on so-called free zone electrophoresis [10, 11]. They described the evolution of separated zones after the application of an electric field and over-loading phenomena. Since this group used PTFE tubes as the separation channel, for detection only a relatively insensitive conductivity detector could be used. Still, the principle of the separation technique was not different from what we now would regard as CE.

The work of Jorgenson and Lukacs, of which an example is shown in Figure 1.2, was astonishing for many in different respects. Those who were used to work with high-performance liquid chromatography (HPLC), were surprised by the narrow peaks in the electropherograms. Instead of the usual 10 000 theoretical plates commonly found as the maximum in HPLC, plate numbers up to 400 000 were shown. Those who were using slab gel electrophoresis, were surprised by the time axis; the separations were performed, not in the usual several hours, but in less than 30 min. Soon, a large number of research groups, at universities and instrument manufacturers, started working on the further instrumental and methodological development of CE. This resulted in an exponentially increasing number of publications on CE in the first half of the nineties. In recent years the research effort on and with CE appears to stabilise. In Figure 1.3 the yearly numbers are given of CE publications as found in a literature search with the help of the Science Citation Index. The biannual fundamental reviews in Analytical Chemistry give the same trend [12–16].

Significant contributions to the success of CE were provided by several groups. Without pretending to be complete, the following landmarks can be listed in the development of CE:

- the introduction of micellar electrokinetic capillary chromatography for the separation of neutral compounds by Terabe c.s. in 1984 [17];

- the introduction of capillary isoelectric focussing for the separation of proteins by Hjerten in 1985 [18];

- the introduction of laser-induced fluorescence (LIF) detection in CE by Gassman et al. in 1985 [19], and by Burton et al. in 1986 [20];

- the development of capillary gel electrophoresis for the separation of DNA fragments and proteins by the group of Karger in the late 1980's [21, 22];

- the introduction of the first complete CE instrument on the market in 1988 [23];

- the development of various methods to reduce the wall adsorption of proteins to be separated by CE [e.g., 24, 25];

- the first successful attempt to couple CE to a mass spectrometer, by means of an electrospray interface, by Smith et al. [26, 27];

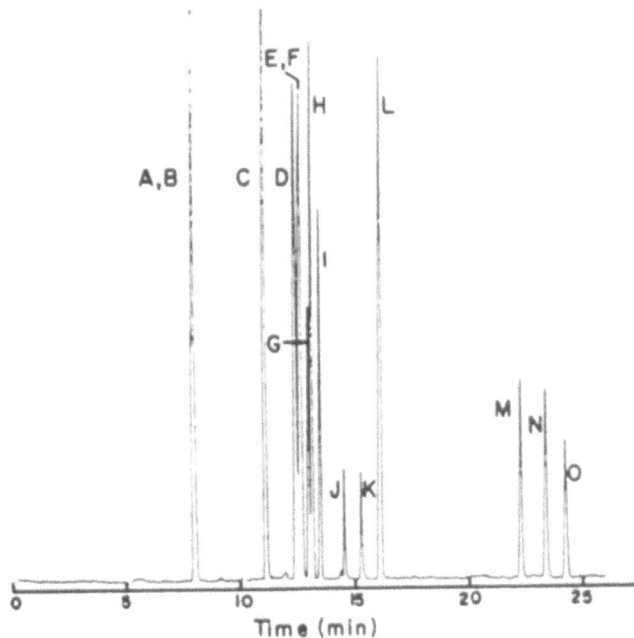

Figure 1.2
Zone electrophoretic separation of dansyl amino acids in a 75 μm Pyrex glass capillary. Reproduced from ref. [1].

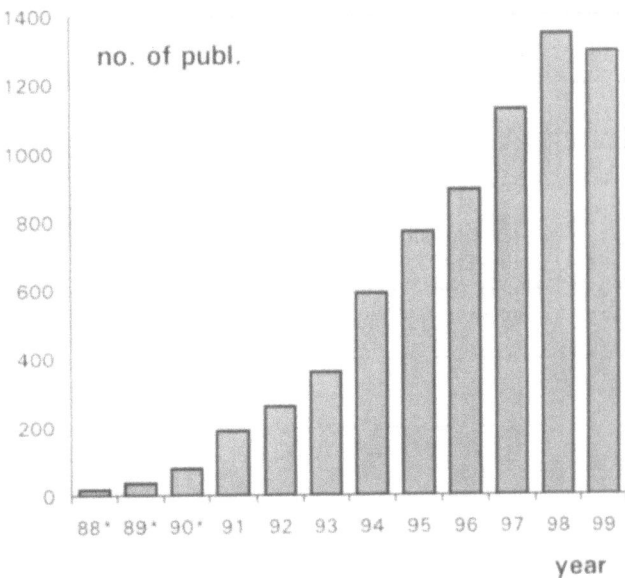

Figure 1.3
Number of publications on capillary electrophoresis, as found with the help of the Science Citation Index. (*: for the years 1988–1990 only publications with the term capillary electrophoresis in the title are counted.)

- the application of CE (with electrochemical detection) for the analysis of the content of a single cell by Ewing c.s. [28];

- the development of capillary electrochromatography (CEC); based on an old idea [29], and already shown casually by Jorgenson and Lukacs in 1981 [30], CEC was brought alive by Knox and Grant [31, 32] and developed into a practical technique by Smith and Evans [33].

At this moment (end of 1999) it seems that CE has developed into a mature, reliable separation method. The main principles, instrumental possiblities and limitations have been sorted out. Research and development work on the main weak point of CE, the detection, is well under way, with LIF and CE-MS coupling as focal points. Still, the utilisation of CE in routine laboratories to solve real analytical problems is not as frequent as might be suggested by the yearly number of research papers on CE, and certainly not up to the expectations of instrument suppliers. After the appearance of a considerable number of suppliers on the market in the last ten years, presently already a shake-out seems to be going on. One possible explanation for the – so far – disappointing commercial success of CE may be a lack of convincing applications. Although in some application areas CE has really brought something new, too many studies have only shown that separations that used to be carried out with HPLC or slab gel electrophoresis can also be realized with CE. Even when it is shown in these cases that CE has certain advantages, in terms of separation speed or efficiency, this may not be enough to persuade a user to discard his/her favorite, well-validated method. The move is with the application chemist.

References

[1] J.W. Jorgenson and K.D. Lukacs, Anal. Chem., 53 (1981) 1298.
[2] F. Kohlrausch, Ann. Phys. Chem., 62 (1897) 14.
[3] A. Tiselius, Trans. Faraday Soc., 33 (1937) 524.
[4] V.P. Dole, J. Am. Chem. Soc., 67 (1945) 1119.
[5] L.G. Longworth, J. Phys. Chem., 51 (1947) 171.
[6] A. Tiselius in "Electrophoresis. Theory, Methods, and Applications", M. Bier (Ed.), Academic Press, New York, 1959, p. xvii.
[7] S. Hjerten, Chromatogr. Rev., 9 (1967) 122.
[8] R. Virtanen, Acta Polytech.Scand., 123 (1974) 1.
[9] F.M. Everaerts, J.L. Beckers and Th.P.E.M. Verheggen, "Isotachophoresis: Theory, Instrumentation, and Applications", Elsevier, Amsterdam, 1976.
[10] F.E.P. Mikkers, F.M. Everaerts and Th.P.E.M. Verheggen, J. Chromatogr., 169 (1979) 1.
[11] F.E.P. Mikkers, F.M. Everaerts and Th.P.E.M. Verheggen, J. Chromatogr., 169 (1979) 11.
[12] W.G. Kuhr, Anal. Chem., 62 (1990) 403R.
[13] W.G. Kuhr and C.A. Monnig, Anal. Chem., 64 (1992) 389R.
[14] C.A. Monnig and R.T. Kennedy, Anal. Chem., 66 (1994) 280R.
[15] R.L.St. Claire, III, Anal. Chem., 68 (1996) 569R.
[16] S.C. Beale, Anal. Chem., 70 (1998) 279R.
[17] S. Terabe, K. Otsuka, K. Ichikawa, A. Tsuchiya, and T. Ando, Anal. Chem., 56 (1984) 111.
[18] S. Hjerten, J. Chromatogr., 347 (1985) 191.
[19] E. Gassman, J.E. Kuo, and R.N. Zare, Science, 230 (1985) 813.
[20] D.E. Burton, M.J. Sepaniak, and M.P. Maskarinec, J. Chromatogr. Sci., 24 (1986) 347.
[21] A.S. Cohen and B.L. Karger, J. Chromatogr., 397 (1987) 409.
[22] A.S. Cohen, A. Paulus, and B.L. Karger, Chromatographia, 24 (1987) 15.
[23] R.G. Brownlee and S. W. Compton, Am. Lab., 20 (1988) 10.
[24] H.H. Lauer and D. McManigill, Anal. Chem., 58 (1986) 166.
[25] G.J.M. Bruin, J.P. Chang, R.H. Kuhlman, K. Zegers, J.C. Kraak, and H. Poppe, J. Chromatogr., 471 (1989) 429.
[26] J.A. Olivares, N.T. Nguyen, C.R. Yonker, and R.D. Smith, Anal. Chem., 59 (1987) 1230.
[27] R.D. Smith, J.A. Olivares, N.T. Nguyen, and H.R. Udseth, Anal. Chem., 60 (1988) 436.
[28] A.G. Ewing, R.A. Wallingford, and T.M. Oefirowicz, Anal. Chem., 61 (1989) 292A.
[29] V. Pretorius, B.J. Hopkins, and J.D. Schieke, J. Chromatogr., 99 (1974) 23.
[30] J.W. Jorgenson and K.D. Lukacs, J. Chromatogr., 218 (1981) 208.
[31] J.H. Knox and I.H. Grant, Chromatographia, 24 (1987) 135.
[32] J.H. Knox and I.H. Grant, Chromatographia 32 (1991) 317.
[33] N.W. Smith and M.B. Evans, Chromatographia, 38 (1994) 649.

2 Basic Principles of CE

2.1 The electromigration of ions

Electrophoresis is the migration of charged particles in a solution under the influence of an electric field. Different particles with different charges and/or sizes migrate with different velocities: this is the basic principle of all electrophoretic separation methods.

The electrostatic force F exerted on a particle i in solution is proportional to the net charge of the particle (q_i) and the electric field strength or voltage gradient (E) in the solution:

$$F = q_i \cdot E \qquad (2.1)$$

Of course the direction of the force is to the electrode with a charge opposite to that of the particle. Under the influence of the electrostatic force the charged particle is accelerated and starts migrating. Its movement is then opposed by viscous forces in the solution, which increase proportional with the velocity v of the particle. For a spherical particle the viscous force is given by the Stokes equation:

$$F = 6\pi\eta \cdot r_i \cdot v_i \qquad (2.2)$$

where η is the viscosity of the solution and r_i the (effective) radius of the particle. After a (very short) acceleration time the opposing forces (electrostatic and viscous) cancel each other out and the particle then moves with a constant velocity through the solution:

$$v_i = \frac{q_i \cdot E}{6\pi\eta \cdot r_i} \qquad (2.3)$$

It is good to realise that under normal conditions in CE, typical migration velocities are in the order of some millimetre per second, while the thermal velocity of simple ions may be several hundreds of meters per second; the electrophoretic migration is nothing more than a small drift in a particular direction superimposed on the stochastic thermal motion of the particles.

For an easier comparison of experimental data obtained with different field strengths, the ionic mobility μ_i has been defined as[a]:

$$\mu_i = \frac{v_i}{E} \qquad (2.4)$$

The dimension of a mobility is $(\text{m s}^{-1}) / (\text{V m}^{-1})$ or m^2 V^{-1} s^{-1}. From the equations above it follows that the mobility of a spherical particle can be written as:

---

[a] Note that in this monograph μ stands for the absolute value of the mobility of a particle.

Table 2.1. Mobilities of inorganic ions.

cations	μ^a	anions	μ^a
H^+	362	OH^-	205
Li^+	40	F^-	57
Na^+	52	Cl^-	79
K^+	76	HCO_3	46
NH_4.	76	NO_3^-	74
Ca^{2+}	62	SO_4^{2-}	83

[a]: absolute values of mobilities [10^{-9} m^2 V^{-1} s^{-1}] in water at infinite dilution, 25 °C.

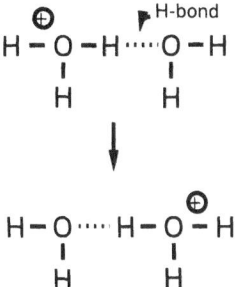

Figure 2.1
Simplified model explaining the high mobility of H^+-ions in aqueous solution.

$$\mu_i = \frac{q_i}{6\pi\eta \cdot r_i} \qquad (2.5)$$

It is clear that different ions can be separated when they differ either in charge (q_i) or in radius (r_i) or, better, when their charge/size ratio differs. The radius determining the mobility is the radius of the moving particle, in aqueous solutions including the hydration shell. The mobilities of a number of simple (inorganic) ions, in water at 25 °C, are given in Table 2.1. The table shows that hydrated radii may differ considerably from the ionic radii in crystals; the mobility of a potassium ion is higher than that of the smaller lithium ion because it binds less water molecules.

The abnormally high mobility of H^+-ions as given in Table 2.1 is not related to the small size of a (bare) proton. In fact, in aqueous solution the radius of a hydrated proton is much larger than that of other simple inorganic ions. With protons in solution another mechanism of charge transport is available, as is illustrated schematically in Figure 2.1. The hydrated proton, for simplicity depicted as a H_3O^+-ion, shows a strong orientating interaction with neighbouring water molecules. When the H_3O^+-ion is properly lined up with the next H_2O-molecule, the excess proton can be virtually moved by a rearrangement of the electronic structure. Since no real mass transport is involved, this process can be much faster than a real ion transport process. Correspondingly, the high mobility of hydroxide ions can be explained as the result of the virtual transport of proton vacancies.

0009-5893/00   S-9-07 $ 03.00/0   © 2000 Friedr. Vieweg & Sohn Verlagsgesellschaft mbH

Effect of the ionic strength

The equations for the mobilities of ions given above are only valid in very dilute solutions, with an ionic strength approaching zero. With a finite salt concentration in the solution, the background electrolyte (BGE) in CE, the mutual interaction of ions will cause a deviating behaviour. In an electrolyte solution an ion is surrounded by other ions. Due to the thermal motions in the solution, any ion may be found at a certain place near the ion in question, but the probability to find an oppositely charged ion is slightly larger because of electrostatic attraction. The excess of oppositely charged ions around a certain ion can be seen as a diffuse ionic cloud, with a charge just neutralising that of the central ion. When the ionic strength of the solution (I) is increased, this ionic cloud will be denser. The diffuse ion cloud around the central ion will migrate in the direction opposite to that of the central ion. Therefore, an extra viscous force is exerted on the central ion, which is stronger for a dense ion cloud (high I). This is called the electrophoretic or retardation effect. Also, when the central ion and the ion cloud have moved in opposite directions, it will take some time for the cloud to rearrange itself around the central ion. During this relaxation time the central ion is 'drawn back' by the electrostatic attraction of the oppositely charged ion cloud. This is called the relaxation effect. Figure 2.2 illustrates the retardation and the relaxation effects. Both effects work in the same direction: the mobility of an ion decreases when the ionic strength of the solution is increased.

The ionic strength effect on the transport properties of (mostly inorganic) ions has been extensively studied in physical chemistry (see for instance ref. [1]). The decrease of the mobilities is found to be roughly proportional to $I^{1/2}$. Friedl et al. [2] conducted a systematic study on the influence of the ionic strength on the mobilities of a large number of sulfonated aromatic compounds. They found that the effect of I could be described to a very good approximation by the empirical expression:

$$\mu/\mu_0 = \exp\left(-0.77\sqrt{z \cdot I}\right) \qquad (2.6)$$

where μ_0 is the mobility in pure water and z the charge number of the ion in question. In Table 2.2 the relative values of mobilities for single and doubly charged ions are given for a range of ionic strength values as usually found in practical CE, calculated with Equation 2.6. As is evident from the table, the effect of the ionic strength for single charged ions is not dramatic. For multiple-charged ions the effect is larger; changing the ionic strength of the BGE may therefore sometimes help to separate overlapping zones of differently charged analytes.

Effect of the pH of the BGE

Clearly, the effective (observed) mobilities of weak acids and bases will depend on the pH of the solution. Since the dissociation equilibrium of a weak acid is very

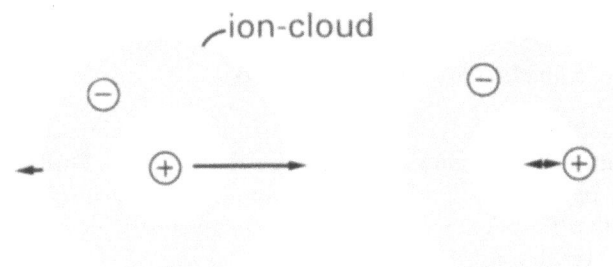

Figure 2.2
Retardation and relaxation effects on ionic mobilities.

Table 2.2. Effect of the ionic strength on the mobilities of ions.

I [mol L^{-1}]	relative mobility $z=1$	$z=2$
0	1	1
0.001	0.98	0.97
0.002	0.97	0.95
0.005	0.95	0.93
0.01	0.93	0.90
0.02	0.91	0.86
0.05	0.84	0.78
0.1	0.78	0.71

fast, every molecule may be regarded as being part of the time in the anionic and part of the time in the neutral form during the separation. Since the electric field exerts a force only on the charged species, the effective mobility of a monobasic weak acid is:

$$\mu_{\text{eff},i} = \alpha_i \cdot \mu_1 \qquad (2.7)$$

where α_i is the degree of dissociation of the acid and μ_1 the ionic mobility of the anion. From the equilibrium equation of a weak acid this gives:

$$\mu_{\text{eff},1} = \frac{K_{a,1}}{K_{a,i} + [H^+]} \cdot \mu_i \qquad (2.8)$$

The corresponding formula for a weak base is:

$$\mu_{\text{eff},i} = \frac{[H^+]}{K_{a,i} + [H^+]} \cdot \mu_i \qquad (2.9)$$

Amphoteric compounds such as amino acids can have a mobility towards the positive or towards the negative electrode, depending on the pH. At low pH, the effective mobility approaches the mobility of the cation, at high pH that of the anionic species. Since the size of the cation and the anion will be approximately equal, these limiting mobilities are expected to be the same. In practice, however, different mobilities for the cationic and the anionic species are sometimes found.

For proteins and other macromolecules the situation is more complicated. Most proteins contain a large number of acidic as well as basic groups. At extreme pH-values of the solution the protein can carry a high positive or negative charge. However, due to the mutual interactions between the different proton-binding sites, there is no sharp change of the net charge with the pH. The mobility of the protein changes gradually with the pH. At the isoelectric point (pI) of the protein the net charge of the molecule, and accordingly its mobility, is zero. Figure 2.3 shows the expected dependency of the mobilities of exemplary compounds on the pH.

The use of additives in the BGE

The complexation with ions present in the BGE can influence the mobility of a species. For instance, in chloride solutions metal ions may migrate towards the positive electrode. In this case the central metal ion is present in solution as a negative chloride complex. This phenomenon should be distinguished from the ion-cloud effect described previously. In complexation, specific short range forces play a role. The influence on the mobility depends on the combination of central ion and complexing ion. The dependency on the ionic strength is a general, non-specific interaction.

In the same way the migration of anions can be influenced by the addition of metal ions to the BGE. For the separation of sugars and catecholamines borate can be added; borate anions complexate with the diol groups of these compounds. Racemic mixtures have been separated by the addition of chiral selectors to the BGE, for instance crown ethers or cyclodextrins. Different effective mobilities result when the two isomers have different complexation constants with the chiral additive.

2.2 Prediction of ionic mobilities

For the design or optimisation of a CE method, it would be of value to know the ionic mobilities of the analytes to be separated *a priori*. Unfortunately, in the vast literature on CE it is not always easy to find a value for the ionic mobility of a simple compound. In original publications on the application of CE, mobilities are often not given; calculation of the appropriate values from the published data is sometimes not possible because essential experimental data are missing. For inorganic ions numerous tables of mobilities have been published. One can also refer to textbooks on physical chemistry or electrochemistry. In these books often tables are given of equivalent ionic conductivity's (λ_0). From these values, ionic mobilities are easily calculated according to:

$$\mu_i = \frac{\lambda_{0,i}}{F} \qquad (2.10)$$

where F is the Faraday constant.

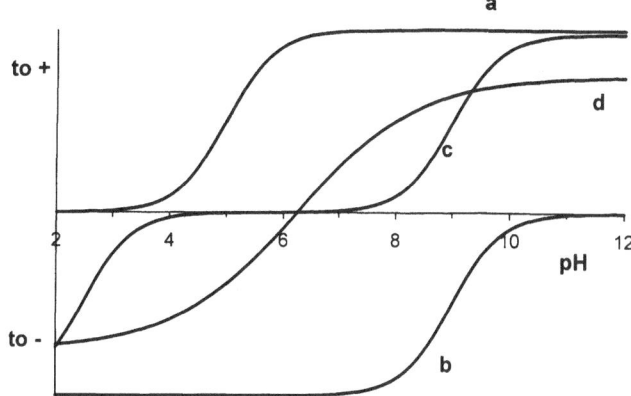

Figure 2.3
Effect of the solution pH on the effective mobilities of weakly acidic and basic compounds. (a) weak acid; (b) weak base; (c) amino acid; (d) protein.

Table 2.3. Mobilities of multiple-charged ions.

ion	mobility[a]			relative mobility		
	$z=1$	$z=2$	$z=3$	$z=1$	$z=2$	$z=3$
phosphate	35.1	61.5	71.5	1	1.75	2.04
citrate	28.7	54.7	74.4	1	1.91	2.59
benzene-tricarbonate	25.1	54.0	84.9	1	2.15	3.38

[a]: mobilities in water at infinite dilution, 25 °C [10^{-9} m^2 V^{-1} s^{-1}]

For organic compounds such tables are not available. A few authors have taken the trouble to compile lists of mobilities of simple organic compounds (e.g., refs. 3, 4, 5). In case the analyte of interest is not listed, one can try to estimate its mobility from the value given for a comparable compound. For such an estimation, a number of rules of thumb can be given. However, such rules should be applied with caution; for every rule numerous exceptions can be found.

Effect of charge

According to Equation 2.5, the mobility of an ion should be proportional to its charge number z. In practice it is generally found that the mobility of a multiply-charged ion is less than z times that of a singly-charged ion of similar size. This phenomenon is also observed for compounds that can have a different charge depending on the pH of the solution. The effect is stronger when the charges are located on the same functional group or close together in the molecular structure, as is shown in Table 2.3. Also, the effect is increased with a higher ionic strength of the solution, because the retardation effect is stronger for higher charged compounds. With ionic charges higher than 3, the proportionality between the charge and the mobility is largely lost.

Effect of molecular weight

Mobilities are expected to be inversely proportional to the size of the ion (Equation 2.5). For homologous series

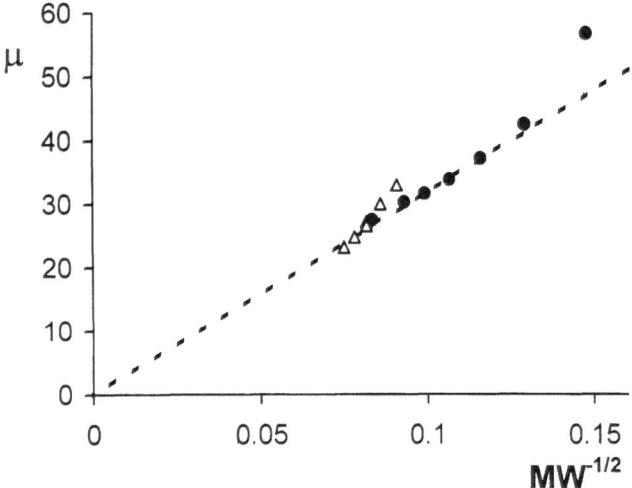

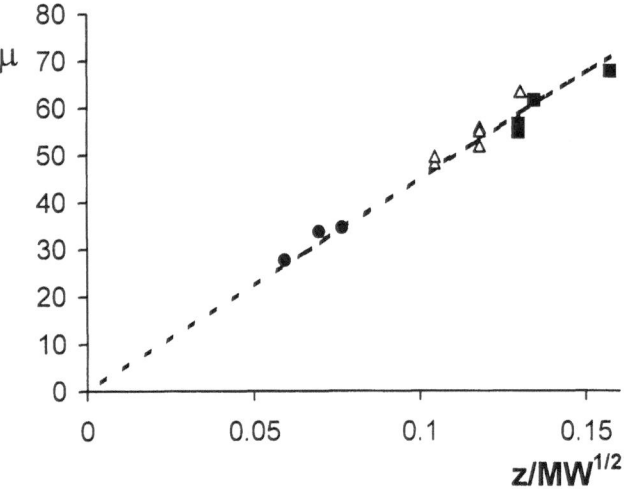

Table 2.4. Mobilities of C_4 dicarboxylic acids

| Dicarboxylic acid | $\mu\,[10^{-9}\,m^2\,V^{-1}\,s^{-1}]$ | |
	$z = 1$	$z = 2$
succinic	35.2	57.5
maleic	42.5	62.0
fumaric	35.1	60.5

Figure 2.4

Dependency of the ionic mobilities of homologous series of compounds on the molecular weight and the charge number. A: singly-charged carboxylic acids (●) and substituted benzoic acids (△). B: aromatic sulfonic acids with z = 1 (●), 2 (△), and 3 (■).

follows that the chemical structure has a decisive influence on the mobility of an ion. This influence is very difficult to predict from the structural formula. Still, a number of general guidelines to predict a mobility can be extracted from data given in the literature.

- The type of the charged group is of importance. Sulfonic acids and sulphates have higher ionic mobilities than the corresponding carboxylic acids, despite their higher molecular weight. Cations often have a higher mobility than anions with the same MW. For amino acids it is sometimes reported that the cationic species has a higher mobility than the anionic species.

- The substitution of extra alkyl groups in the structure decreases the mobility of an ion approximately according to the MW-rule. The decrease by the substitution of an aryl group is usually less than may be expected on basis of its contribution to the MW.

- The effect of the substitution of a hydroxyl-group is either small, or it may even be that the ionic mobility is increased. The mobility of α-hydroxybutyrate is higher than that of butyrate; that of salicylate higher than that of benzoate. Other hydrophilic groups (keto, non-ionised carboxylic acid, amide groups) have often only a small effect.

- The substitution of hydrophobic groups, such as nitro-groups or chlorine or bromine atoms, gives a decrease of the mobilities which is usually smaller than that caused by a simple methyl-group.

As stated before, these guidelines should be applied with reserve. As an illustration Table 2.4 gives the reported mobilities for the three dicarboxylic acids with 4 carbon atoms: succinic acid (the saturated compound), malice acid (the cis-isomer) and fumaric acid (the transisomer of the unsaturated compound). Not only large differences in mobilities are shown, these differences also depend on the charge number of the ions.

Mobilities of peptides

For certain classes of compounds empirical expressions have been formulated that correlate observed ionic mobilities with structural data. Most of this kind of work has been carried out for peptides [6, 7, 8]. The formulas give the expected mobility as a function of the (calculated) effective charge of a peptide under certain experimental conditions, and the number of amino acid units or the molecular weight. For the latter factor (n or MW), exponents between −0.3 and −0.7 have been found.

or groups of compounds with similar structure, the size can be conveniently expressed by the molecular weight (MW) of the compound of interest. The mobilities within a series are inversely proportional to the square root of the MW. Examples of this relation are shown in Figure 2.4A. Except for formic acid, the relation holds well for normal carboxylic acids. Alkyl substituted benzoic acids also follow the rule, although their mobilities are relatively higher than those of the alkyl carboxylic acids. In Figure 2.4B mobilities are plotted for aromatic compounds with multiple sulfonic acid groups, as measured by Friedl et al. [2], as a function of $z/MW^{1/2}$. Up to a charge number of three a good linear relation is found; compounds with a higher number of sulfonic acid groups did not fit well in the plot.

Effect of substituents

From the fact that compounds with the same charge and similar molecular weight are often easy to separate, it

Basic Principles of CE

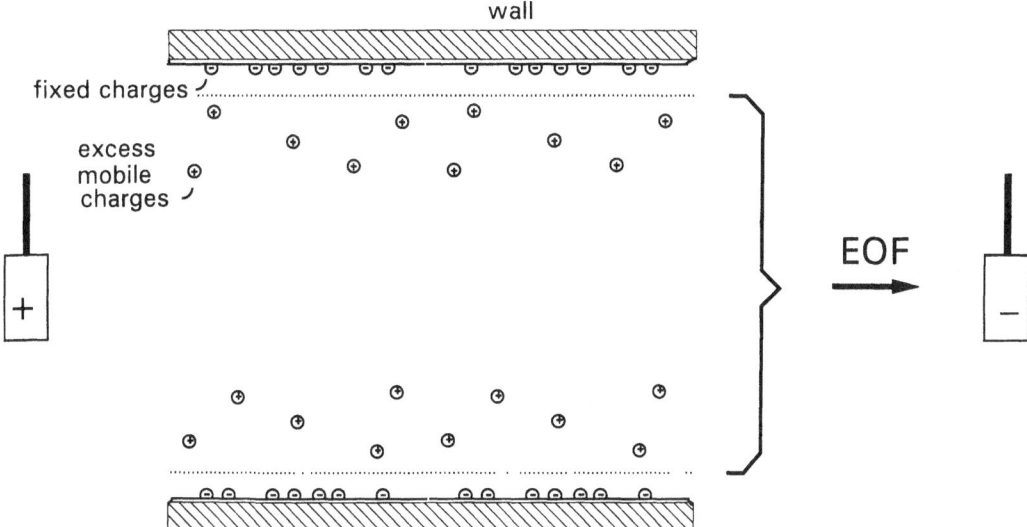

Figure 2.5
The principle of electroosmosis.

2.3 The principles of electroosmosis

When an aqueous solution of electrolytes, as used in electrophoresis, is in contact with the wall of the separation capillary, there is a charge separation between the wall and the solution. This can be caused by ionisation of the wall material or by specific adsorption of ions from the solution to the wall. With fused silica capillaries the wall is usually negatively charged. Free silanol groups on the surface of the fused silica are deprotonated (at pH > 1.5), leaving negative Si-O⁻ groups. Since the system as a whole must be electrically neutral, the solution in the separation compartment has a net positive charge. This excess of positive ions is located in the solution close to the wall, due to the electrostatic attraction by the negative wall. In Figure 2.5 this charge separation in a capillary is shown schematically. When a voltage is applied between the ends of the capillary, the electric field exerts a force on the excess of positive charge in the solution close to the wall. This force drives the solution in the capillary as a whole in the direction of the negative electrode. A constant flow of the solution results when the viscous forces in the thin layer of solution near the wall counteract the electrostatic force. This phenomenon is called electro-osmosis or electro-endoosmosis. Electro-osmosis is an important feature of CE since the separation is performed in free solution. In classical electrophoresis with its highly viscous gels or agars it generally does not play a role.

The velocity of the electroosmotic flow (EOF) is proportional to the field strength. An electroosmotic mobility can be defined similar to an ionic mobility:

$$\mu_{eo} = \frac{v_{eo}}{E} \tag{2.11}$$

The unit of the electroosmotic mobility is also m² V⁻¹ s⁻¹. It depends mainly on the wall material and the pH of the solution (who together determine the charge density on the wall) and on the viscosity of the solution. For a fused silica capillary at pH 7 it is in the order of $50\text{–}75 \cdot 10^{-8}$ m² V⁻¹ s⁻¹.

Often the osmotic mobility is larger than the ionic mobilities of the analytes in the solution. This implies that all solutes migrate towards the negative electrode, with a velocity reflecting an apparent mobility (μ_{app}):

- for positive ions: $v_i = (\mu_{eo} + \mu_{eff,i}) \cdot E = \mu_{app,i} \cdot E$
- for negative ions: $v_i = (\mu_{eo} - \mu_{eff,i}) \cdot E = \mu_{app,i} \cdot E$
- for neutral particles: $v_i = \mu_{eo} \cdot E$

When the sample is introduced at the positive end of the capillary and the detector positioned at the negative end, all solutes can be determined in one run. From the net velocity of an ion, as obtained from the length of the capillary up to the detector position (L_d) and the migration time (t_i), the apparent mobility can be calculated:

$$\mu_{app,i} = \frac{L_d}{t_i \cdot E} \tag{2.12}$$

In Table 2.5 the different definitions of mobilities, as used throughout this book, are summarised. Note that in this book any μ stands for the absolute value of a mobility; where necessary the direction is indicated by a + or − sign.

An important characteristic of the EOF is that the flow profile in the capillary is flat. Since the electrostatic force acts only on the outer layer of the solution, with its excess of positive ions, the whole solution moves with the same speed, except for a very thin layer (in the order of a few nm) close to the wall. This implies that the EOF does not cause broadening of zones in CE.

The EOF and the effects of the capillary wall properties and the BGE composition on it will be further discussed in Chapter 6.

Table 2.5 Definitions of the mobility as used in this monograph

symbol	name	definition
μ	ionic mobility	mobility of a specific charged species of a compound, relative to the solution
μ_{eff}	effective mobility	average mobility of the different species of a compound that are in rapid equilibrium, relative to the solution
μ_{app}	apparent mobility	effective mobility of a compound, relative to the capillary

2.4 Separation modes in CE

CE can be regarded as the collective name for a number of electrophoretic separation techniques. These different separation modes have only in common that they are performed in a narrow capillary with a high-voltage source providing the energy required for the separation. Still, all subtechniques can be carried out with the same instrumentation; the difference is determined by the choice of the separation media and the type of capillary.

An extensive discussion of the principles, implementation and merits of the various separation modes in CE is outside the scope of this book. Here, only their essential characteristics will be summarised briefly.

Capillary zone electrophoresis (CZE)

Capillary zone or free zone electrophoresis is the most common separation mode in CE. In CZE, the separation of analyte ions is carried out in a background electrolyte (BGE) present in the capillary and in both electrode vials before the analysis. The sample solution is introduced as a small plug on one side of the capillary. After application of the voltage, different zones of analyte ions are developed from the sample plug that migrate with different velocities towards the detector side of the capillary. The BGE usually consists of an aqueous solution of an acid-base buffer. The function of the BGE is to keep the electrophoretic conditions as constant as possible over the capillary length during the separation. Ideally, the presence of analyte ions in a zone has a negligible influence on the local field strength and solution properties. The migration time of a zone, the time it takes the zone to migrate to the detector, is taken as characteristic for the particular sample ion; the height or area of the peak-shaped detector signal is a measure for the concentration of the compound in question in the sample solution.

Capillary isotachophoresis (CITP)

CITP is the capillary variant of the almost classical isotachophoretic separation method [9]. ITP resembles Tiselius' moving boundary electrophoresis; it is carried out in a discontinuous electrolyte system. The separation capillary and the end-side electrode vial are filled with a leading electrolyte before the separation, the injection-side vial with a terminating electrolyte. The leading electrolyte contains an ion type with a higher mobility than any of the analyte ions of interest; the mobility of the terminating electrolyte ion is lower than that of the analyte ions (with the same sign). The sample solution is stacked, as a small plug, between these two different electrolytes at the start of the separation. Because of the properties of the electrolytes, the analyte ions are trapped between the leading and terminating ions. During electrophoresis the sample ions are separated into adjacent zones. The composition of these zones is completely determined by the composition of the leading electrolyte; therefore, the length of an analyte zone is a measure for the amount of a type of ion in the injected sample. Zones are identified by their detection properties (conductivity or absorptivity).

Capillary gel electrophoresis (CGE)

Macromolecular polyelectrolytes such as DNA fragments can not be separated by size with CZE. Because the charge/size ratio of molecules of the same type with different size is almost constant, their mobilities will be approximately equal. Therefore, an additional separation mechanism is required. In CGE, as in slab gel electrophoresis, this is the sieving effect of the polymeric network of a gel. Smaller molecules are less hindered by the network in their migration than larger molecules are, so that their mobility is higher.

In the beginning of CGE, the same cross-linked gels were used as the separation medium as in slab gel electrophoresis. The preparation and handling of gel-filled capillaries is not easy; moreover, it appeared that a particular capillary could be used for only a limited number of analyses. Therefore, the introduction of the use of so-called entangled polymer solutions in CE, by Chin and Colburn [10] and by Zhu et al. [11] in 1989, was of much practical importance. For an entangled polymer solution, linear, non-crosslinked polymers of, e.g., hydroxyethyl cellulose (HEC) are used. It appeared that the mesh formed by the polymer molecules in such a solution provides a similar sieving effect as a cross-linked gel. The advantage is that the viscosity of the solution can be low enough to be able to pump it in and out of the capillary. Special techniques for the preparation of the gel-filled capillary are not necessary anymore, and for every analysis the polymer solution in the capillary can be refreshed.

CGE is mostly used for DNA analysis. Another important application is for the separation of proteins as (denatured) SDS complexes. Since the charge/size ratio of SDS-protein complexes is also almost constant, by the sieving effect of the gel a separation according to molecular weight can be accomplished.

Capillary isoelectric focusing (CIEF)

Isoelectric focusing of proteins, well-known in its slab gel variant, can also be performed in a capillary [12].

The pH-gradient in the capillary is formed electrophoretically, by passing current, with a low pH solution in the positive electrode vial and a high pH solution in the negative electrode vial. CIEF can be carried out with focusing and detection in one sequence, or as two different steps in the procedure. In the first case, an electroosmotic flow is required. The analytes need a certain time (capillary length) for focusing, and are subsequently swept through the detection zone by the EOF. In the second case, focusing is carried out with a suppressed EOF. After focusing, the zones are mobilised, by applying a hydrostatic pressure or by electrophoretic means. During mobilisation the sharp zone boundaries can be preserved by maintaining the electric field. In a number of cases it has been shown that the separation efficiency of CIEF is higher than that of the slab gel variant.

Micellar electrokinetic capillary chromatography (MECC)

MECC (also sometimes abbreviated as MEKC) has been introduced by Terabe et al. [13] for the separation of neutral compounds. In respect to operation, the only difference with CZE is that an ionic surfactant is added to the BGE. The principle of the separation, however, is completely different. In MECC, a separation is based on differences in the distribution of compounds between two phases. MECC is therefore not an electrophoretic but a chromatographic technique. The aqueous BGE constitutes the mobile phase, while charged micelles of the surfactant act as the (pseudo-)stationary phase. Neutral analytes, distributed between the two phases, migrate with a velocity between that of the mobile phase (the electroosmotic velocity) and that of the micelles. MECC has become one of the most successful subtechniques of CE. The addition of a micelle-forming surfactant to the BGE has even become routine for the separation of charged compounds, in order to add extra selectivity to the system.

Capillary electrochromatography (CEC)

CEC is a chromatographic separation technique. In CEC a capillary is used packed with stationary phase particles. The high-voltage source of a CE set-up is merely used to drive the mobile phase through the column, by means of the electroosmotic process. The advantage of CEC over pressure-driven chromatography is that there is no pressure limit on the size of the particles that can be used. With the very small particles that can be used in CEC (down to 1 μm), high plate numbers can be achieved. The main practical problem with CEC seemed to be in the preparation of the frits at the ends of the column packing. Since this problem has been solved [14], CEC can be carried out with a standard CE instrument after minor modifications.

References

[1] H.S. Harned and B.B. Owen, "The Physical Chemistry of Electrolytic Solutions", Reinhold Publ.Corp., New York, 1958.
[2] W. Friedl, J.C. Reijenga, and E. Kenndler, J. Chromatogr., 709 (1995) 163.
[3] T. Hirokawa, M. Nishino, and Y. Kiso, J. Chromatogr., 252 (1982) 49.
[4] T. Hirokawa, S. Kobayashi, and Y. Kiso, J. Chromatogr., 318 (1985) 195.
[5] J. Pospichal, P. Gebauer, and P. Bocek, Chem. Rev., 89 (1989) 419.
[6] P.D. Grossman, J.C. Colburn, and H.H. Lauer, Anal. Biochem., 179 (1989) 28.
[7] B.J. Compton, J. Chromatogr., 559 (1991) 357.
[8] A. Cifuentes and H. Poppe, J. Chromatogr. A, 680 (1994) 321.
[9] F.M. Everaerts, J.L. Beckers and Th. P.E.M. Verheggen, "Isotachophoresis: Theory, Instrumentation, and Applications", Elsevier, Amsterdam, 1976.
[10] A.M. Chin and J.C. Colburn, Am. Biotech. Lab./News Ed., 7 (1989) 10 A.
[11] M. Zhu, D.L. Hansen, S. Burd, and F. Gannon, J. Chromatogr., 480 (1989) 311.
[12] S. Hjerten, J. Chromatogr., 347 (1985) 191.
[13] S. Terabe, K. Otsuka, K. Ichikawa, A. Tsuchiya, and T. Ando, Anal. Chem., 56 (1984) 111.
[14] N.W. Smith and M.B. Evans, Chromatographia, 38 (1994) 649.

3 Efficiency and Resolution

3.1 Sources of zone broadening

The terminology to express the separation performance in CE is largely copied from chromatography. Peak widths in the electropherograms are often given by the standard deviation (σ) or variance (σ^2) of the peaks, the efficiency of the system by the observed plate number (N), and the degree of separation of two adjacent peaks by their resolution (R_s). Occasionally the similarity between chromatography and CE is overestimated, when an efficiency is expressed as the number of plates per meter. As will be shown below, plate numbers in CE do not increase proportionally with the capillary length.

In most variants of CE, the sample is introduced as a very short plug at the tip of the capillary. During electrophoresis the resulting analyte zones get broader by a number of causes. Independent contributions to zone broadening are added as variances:

$$\sigma_{tot}^2 = \sigma_1^2 + \sigma_2^2 + \sigma_3^2 + \ldots \tag{3.1}$$

This implies that the total width of the resulting zone when it arrives at the detector position, is dominated by the largest contribution to zone broadening. Improving the efficiency of a separation therefore starts with finding this main source.

The standard deviation of a zone can be expressed in units of length in the direction of the axis of the capillary (σ_l), or in units of time in the electropherogram (σ_t). The plate number is then calculated as:

$$N = \frac{L_{det}^2}{\sigma_l^2} = \frac{t_i^2}{\sigma_t^2} \tag{3.2}$$

where L is the length of the capillary up to the detector and t_i the migration time of the analyte zone. The main contributions to zone broadening will be discussed in some detail.

Axial diffusion

A sharp zone becomes wider by diffusion of the analyte molecules in the direction of the axis of the capillary. This process causes the zone width to increase with the square root of the time (see Figure 3.1), or the variance of the zone proportional with time:

$$\sigma_{dif}^2 = 2\,D.t \tag{3.3}$$

where the standard deviation σ is expressed in units of length. In this equation D is the diffusion coefficient of the analyte ion. The broadening of zones by axial diffusion does not rely on the application of the voltage. Therefore, the waiting time, e.g., between sample injection and the separation or for the changing of vials for fraction collection, should be kept as short as possible. For a small inorganic ion the diffusion coefficient D is typically in the order of 10^{-9} m^2 s^{-1}. From an infinitely

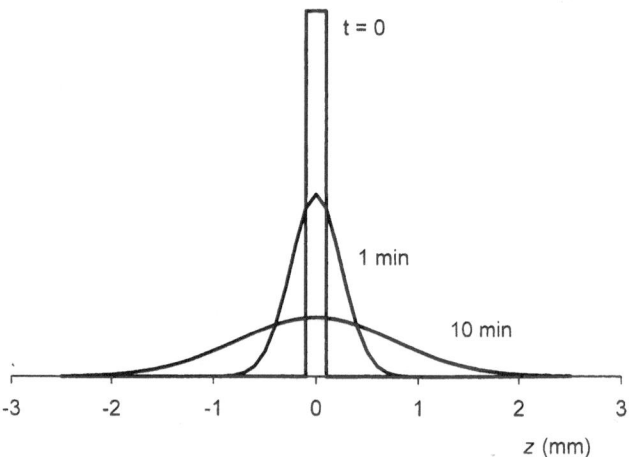

Figure 3.1
Broadening of a zone by diffusion in the axial direction.

narrow sample plug a zone will develop in, e.g., 500 s with a standard deviation of 1 mm.

Ideally, axial diffusion is the only source of zone broadening left, and the ultimate efficiency of CE is determined by it. Of importance is the width of a zone at the time it reaches the detector (t_i). Increasing the speed of the separation by increasing the applied voltage will decrease the zone broadening by axial diffusion. The migration time of an analyte ion is given by:

$$t_i = \frac{\mu_{app,i} \cdot E}{L_{det}} \approx \frac{\mu_{app,i} \cdot V_{appl}}{L^2} \tag{3.4}$$

where V_{appl} is the applied voltage and L the length of the capillary. (In Equation 3.4 it is assumed that the detector position is close to the end of the capillary). Since the plate number is equal to L^2/σ^2 when σ is in length units, we find for the ultimate plate number:

$$N = \frac{\mu_{app,i} \cdot V_{appl}}{2D} \tag{3.5}$$

The advantage of using narrow capillaries for the separation efficiency of electrophoresis is directly related to the higher voltages that can be applied. For small ions $\mu_{app,i}/2D$ is in the order of 10–20 V^{-1}; with applied voltages up to 35 kV plate numbers over 500 000 are possible. Large molecules have smaller diffusion coefficients. Therefore, the contribution to zone broadening by diffusion is smaller, and in CE the highest efficiencies are obtained with macromolecules. Plate numbers of several millions are very well possible.

As is shown in Equation 3.5, the ultimate plate number does not depend on the length of the capillary. By using shorter capillaries the speed of the analysis can be increased without compromising the separation efficiency. However, as will be shown below, there is a trade-in of the sensitivity because the sample volume that can be introduced is proportional to the capillary length.

0009-5893/00   S-16-04 $ 03.00/0   © 2000 Friedr. Vieweg & Sohn Verlagsgesellschaft mbH

Injection and detection

After introduction of the sample the zones already have a certain length. As will be discussed elsewhere, the zones may get broader or sharper immediately after the start of electrophoresis, depending on the conductivity's of the sample and the separation buffer. However, neglecting this special effect, the contribution to the total zone broadening (σ_l) is proportional to the length L_{inj} of the sample plug, with a proportionality constant, depending on the method of sample injection, between 0.3 and 0.5 (see Chapter 8). In cases when one wants to obtain the highest sensitivities, the maximum length of the sample plug that can be introduced is easily calculated from the capillary length and the plate number that is required to obtain a satisfactory separation, by means of Equation 3.2.

Detection may also contribute to zone broadening. With UV-absorbance or fluorescence detection, a certain length (L_{det}) of the capillary is used as the detector 'cell'. Even the narrowest zones give a signal of a certain width while they migrate over the detection window, with a σ_l contribution equal to 0.3 times the window length.

Effects of Joule heating

During electrophoresis heat is produced in the solution by the passage of the current through it. The heat is produced homogeneously in the solution, while it can dissipate only through the capillary wall. Apart from a general rise of the temperature of the solution, also a temperature gradient in the solution inside the capillary is developed. In the centre of the capillary the temperature is higher than close to the wall. This makes the solution in the centre less viscous, and the mobility of the analyte ions higher. Therefore, the ions migrate faster in the centre of the capillary than near the wall. This broadening effect is counteracted by diffusion in the radial direction: ions near the wall lagging behind may diffuse to the centre where they can catch up with the zone, and vice versa (see Figure 3.2). For the zone broadening by temperature gradients the following can be stated [1, 2, 3]:

- it increases strongly with the applied field strength E;
- it increases strongly with the diameter of the capillary; in wide capillaries the temperature gradients are stronger, and the counteracting effect of radial diffusion is smaller;
- it increases with the current, or the conductivity κ of the buffer solution;
- the effect is stronger for ions with a high (positive or negative) mobility; the neutral zone is not affected.

In modern CE this zone broadening contribution is observed only at extremely high voltages. As long as the capillary diameter is $100\,\mu m$ or less, and the conductivity of the BGE is not extremely high ($\kappa < 1\,\Omega^{-1}\,m^{-1}$), the effect on the zones is negligible for small ions. For the separation of macromolecules radial temperature

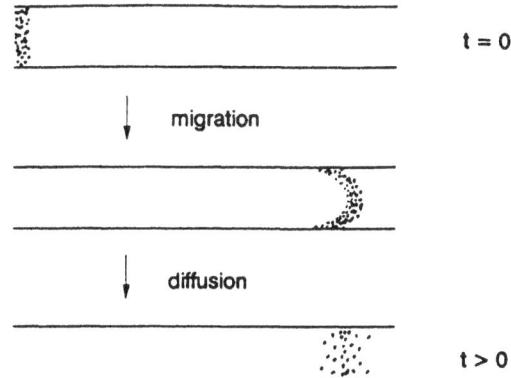

Figure 3.2
Broadening of a zone by a radial temperature gradient, counteracted by diffusion in the radial direction.

gradient effects are of more importance. In the first place, the counteracting effect of radial diffusion is less effective because of the lower diffusion coefficients. Secondly, higher plate numbers are expected *a priori* for such compounds, so that any zone broadening contribution is easier noticeable.

Still, even when in theory the effect of temperature gradients is negligible, in practice it is often found that the plate number drops when the electrophoretic current exceeds a certain limit. Part of the loss of efficiency has been explained as the result of a rise of the average temperature of the solution [4]. With insufficient thermostating the BGE temperature is (locally) easily elevated close to its boiling point. At higher temperature the viscosity of the BGE is decreased. This has a stronger effect on diffusion coefficients than on mobilities, so that the ratio of μ_{app}/D, and with that the plate number (see Equation 3.5), will be decreased.

Another factor contributing to the drop of the plate number at high currents may be non-uniform thermostating. Temperature differences over the length of the capillary do not cause problems in this respect. At a part of the capillary where the (average) temperature is higher than elsewhere, the effect of the increased (electroosmotic as well as ionic) mobilities is exactly compensated by a decreased local field strength. However, when locally one side of the capillary is less efficiently cooled than the opposite side, a difference in the analyte velocity over the cross section of the capillary will result, and zones will become broadened.

Hydrodynamic flow

When the solution level in one electrode vial is higher than in the other, a flow of solution will result by siphoning. This flow is laminar, i.e., it flows faster in the centre of the capillary than close to the walls. Thereby, zones will be 'smeared out'. Siphoning is stronger in wide capillaries. With diameters of $50\,\mu m$ or less, it is usually not a problem [5]. With $\geq 100\,\mu m$ capillaries the position of the vials and the volume of liquid in them should be carefully controlled; a height difference of a few millimetres may already cause an appreciable broadening. The effect is stronger for large molecules, since for these the counteracting radial diffusion is less effective.

Hydrodynamic flow can also result from differences in the electroosmotic velocity over the length of the capillary. Since the total liquid flow must be constant, a difference in the EOF between two parts of the capillary will be compensated by a superimposed (laminar) hydrodynamic flow (see Figure 3.3). Such differences in the electroosmotic mobility may be caused by differences in the surface properties of the inner wall of the capillary. Also, they can result from a difference in the pH of the solution over the capillary length. One should be especially aware of this possibility when the large volume of sample is injected, for instance when the sample stacking technique is applied (see Chapter 8). Similar problems can be encountered when off-column detection (e.g., with electrochemical detection or MS coupling) is applied [6].

Adsorption of analytes to the wall

When the analyte is adsorbed to the wall this usually ruins the separation efficiency. Even when the adsorption is weak and migration times are hardly affected, the plate number drops dramatically [7]. This is shown in Figure 3.4, where the expected plate number is given as a function of the capacity factor for the adsorption process. As in reversed phase HPLC, the adsorption of amine compounds on a surface with free silanol groups (the fused silica wall is full of them) is notorious. A detailed treatment of the effect of adsorption is not very useful here: to preserve the efficiency of the separation adsorption should be avoided. Methods to decrease wall adsorption will be discussed in Chapter 6.

Comparison of sources

Of the various zone broadening sources discussed above, some are independent of the applied voltage (injection and detector contributions), some decrease with the field strength E (axial diffusion and siphoning effects) and some increase with E (temperature gradients and adsorption effects). In principle, an optimum value for the field strength can be found, at which the plate number is at its maximum. In Figure 3.5 plots are given of the plate height H (L/N) as a function of the applied field strength, calculated for the separation of small ions under typical CE conditions. The plots have been constructed for capillary diameters of 100 and 200 μm. It is shown that:

- with a narrow capillary lower plate heights (and higher plate numbers) can be obtained;
- with narrow capillaries a higher field strength can be applied before temperature gradients start to play a role, so that a higher separation speed is obtained.

These two points describe exactly the advantage of CE over classical electrophoretic methods.

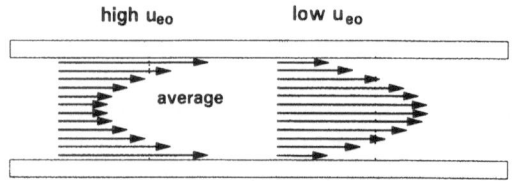

Figure 3.3
Hydrodynamic flow profiles superimposed over the EOF when differences in the electroosmotic mobility exist over the length of the capillary.

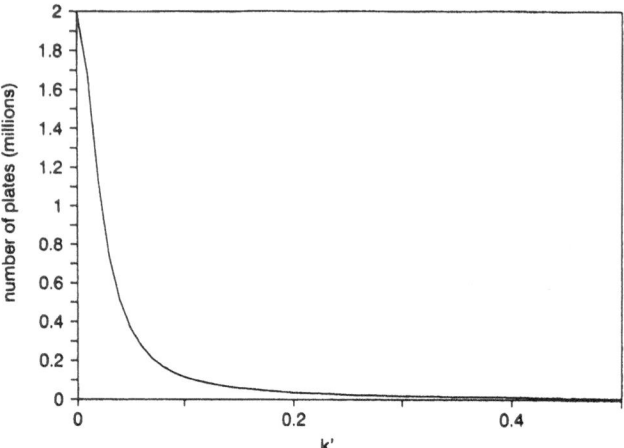

Figure 3.4
Effect of the strength of adsorption on the wall of the capillary, indicated as the capacity factor k', on the observed plate number. Capillary diameter: 50 μm; $v_{app} = 0.5$ mm s^{-1}; $D = 10^{-10}$ m^2 s^{-1}.

In Table 3.1 the various zone broadening contributions are compared to be encountered in a typical CE separation. The analytes considered are benzoic acid as a typical small ion and albumin as a typical protein, separated in a 0.05 mol L^{-1} phosphate buffer at pH = 7, with a capillary of 75 μm diameter, a length of 0.6 m ($L_d = 0.5$) and an applied voltage of 30 kV. A height difference of 5 mm is assumed between the vials. Although the numbers given in the table are only meant as indicative, it can be seen that for small ions axial diffusion is the limiting factor. When separating macromolecules, the constraints on the sample volume and detection window length are more severe; also, siphoning should be more strictly avoided.

3.2 Resolution

Often for small ions the main contribution to zone-broadening is the axial diffusion. The plate number N is then given by Equation 3.5. In this case the plate number is proportional to the available voltage. However, this limiting situation is only met:

- when the sample amount injected is very small;
- when the detector 'window' is very narrow;
- when the diameter of the capillary is small;
- when adsorption can be avoided.

From Equation 3.5 it seems that a high electroosmotic flow is advantageous for the separation. However, even

Efficiency and Resolution

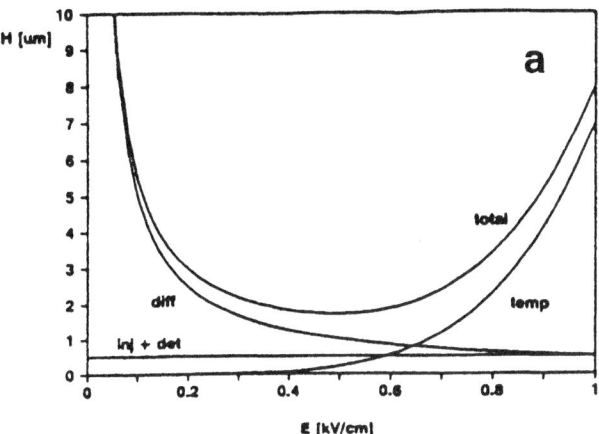

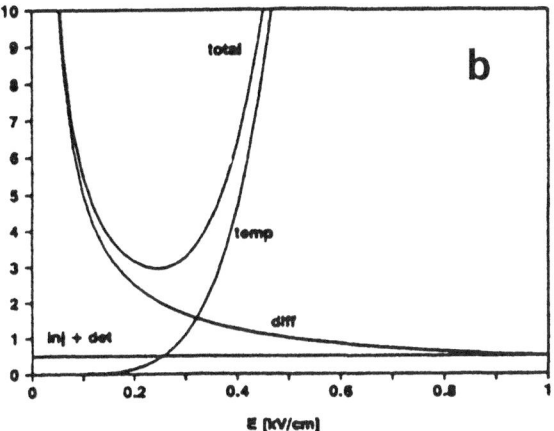

Figure 3.5
Typical dependency of the plate height (L/N) contributions on the applied field strength in CE. (a) narrow capillary, (b) wide capillary.

Table 3.1 Typical zone broadening contributions in CE.

Zone broadening source	σ_l [mm]	
	benzoate	albumin
axial diffusion	1.00	0.10
temperature gradient	0.17	0.14
injection and detection	0.30	0.30
siphoning	0.08	0.25
total	1.06	0.43
N	220,000	1,350,000

when the plate number increases with μ_{eo}, the resolution between two adjacent peaks deteriorates. The resolution between two peaks i and j is defined in the same way as in chromatography:

$$R = \frac{t_j - t_i}{4\sigma_t} \tag{3.6}$$

For two closely spaced peaks, with a mobility ratio α ($\alpha > 1$), the resolution is:

$$R = \frac{1}{4}\,(\alpha - 1)\,\frac{\mu_i}{(\mu_{eo} \pm \mu_i)^{1/2}}\left(\frac{V_{app}}{2D}\right)^{1/2} \tag{3.7}$$

Table 3.2 Comparison of separation performances of CE and HPLC.

separation method	plate number	minimum α^a	peak capacity
CE[b] for cations	1,500,000	1.008	140
CE[b] for anions	300,000	1.004	200
isocratic HPLC	10,000	1.04	75

[a]: for $R = 1$; [b]: $V = 30\,\text{kV}$, $\mu_{eo} = 60$; $\mu_i = 0\text{–}40$; $D = 10^{-9}$.

The best resolution is obtained when the analyte ions migrate against the electroosmotic flow, and ($\mu_{eo} - \mu_i$) approaches zero. This approach has been applied for instance for the electrophoretic separation of ^{35}Cl and ^{37}Cl ions; a baseline separation could be obtained by careful tuning of the electroosmotic velocity [8]. With a high electroosmotic flow in the same direction as the migration, the length of the capillary and hence the voltage is used ineffectively; a larger part of the capillary is just necessary to accommodate the electroosmotic flow. From Equation 3.7 one can also calculate the minimum mobility ratio that is required to obtain a certain resolution between two peaks. It appears that it takes a larger difference in mobilities to obtain a complete separation for compounds migrating in the same direction as the EOF than for compounds migrating against it (see Table 3.2).

Another way to look at the efficiency of separation methods is by comparing peak capacities. The peak capacity is a way to express the information content of an electropherogram (see also ref. [9]). The peak capacity is defined as the number of peaks that in theory can be separated from each other with a resolution of one between each peak pair. Of course in a real separation it will be impossible to find conditions where the peaks are spaced so ideally. Still, the peak capacity of a method gives a good impression of its ability to separate complex mixtures. In Table 3.2 typical values of the peak capacity of CE for cations (assumed to migrate with the EOF) and for anions (migrating against the EOF) are given. For comparison, data are also included for an isocratic HPLC system. As expected the peak capacity of CE is superior, although the difference may be smaller than expected on basis of the typical plate numbers.

References

[1] R. Virtanen, Acta Polytech. Scand., 123 (1974) 1.
[2] J.H. Knox and I.H. Grant, Chromatographia, 24 (1987) 135.
[3] W.Th. Kok, Chromatographia, 24 (1987) 442.
[4] J.A. Knox and K.A. McCormack, Chromatographia, 34 (1994) 207.
[5] G. Roberts, P. Rhodes, and R. Snyder, J. Chromatogr., 480 (1980) 35.
[6] W.Th. Kok, Anal. Chem., 65 (1995) 1853.
[7] G.J.M. Bruin, J.P. Chang, R.H. Kuhlman, K. Zegers, J.C. Kraak, and H. Poppe, J. Chromatogr., 471 (1989) 429.
[8] C.A. Lucy and T.L. McDonald, Anal. Chem., 67 (1995) 1074.
[9] E. Kenndler, Chromatographia, 30 (1990) 713.

Efficiency and Resolution

4 Voltages and Currents

4.1 The voltage source

As has been shown in Chapter 3, the separation efficiency that can be obtained in CE is proportional to the voltage that is applied. A high-voltage source is therefore a prerequisite for CE. Commercial instruments have sources with an upper limit of usually 30 or 35 kV. With higher voltages the electrical insulation of the various parts of the CE set-up becomes problematic, with spontaneous sparking and a risk of serious damage of sensitive electronic components as a result. A high precision and low drift of the applied voltage is of course important for the reproducibility of the migration times of analytes.

Some instruments have the option to apply a constant current rather than a constant voltage. In some cases this gives a better reproducibility [1, 2]. This is certainly the case when the capillary is not properly thermostatted. When the temperature of the solutions is increased, the mobilities of ions are generally increased with 2–3 % per °C. With a constant voltage this implies a decrease of the retention times with 2–3 % per °C. However, not only the mobilities of the analytes are affected, but also those of the components of the background electrolyte, and consequently the conductivity of the solution. With a constant-current source this means that automatically the voltage will be reduced when the temperature of the solution is increased. Therefore, the retention times are then hardly affected by a change of the temperature in the system.

In most cases a ripple or high-frequency noise on the output of the voltage source is of no importance. An exception is found with electrochemical or conductivity detection. Here, the noise on the high voltage may be transferred into a noise on the detector signal. It has also been reported that the noise on the signal of a UV detector strongly increased when the high voltage was applied. This was explained as the result of vibrations of the capillary in the holder of the optical system. The use of a power supply with a lower ripple on the voltage will reduce such problems. With high quality voltage sources the ripple can be as low as a few volts on 30 kV.

An important aspect for the practising chemist is the safety of the instrumentation used. The high voltage source should be provided with a properly working ground-leak detector. Since the chance of ground-leakage is the largest at the detector side of the capillary (where the end of the capillary is closest to other instrument parts), it is common practice to ground the voltage source at this electrode side. Nevertheless it is advisable when home-made instrumentation is used, to place the whole instrument in a protective cabinet with a contact that rapidly switches off the power when the cabinet is opened.

4.2 The field strength and electrophoretic current

In capillary electrophoresis a voltage of several kilovolts is applied over the length of the capillary. The migration velocity of an ion at a specific point in the capillary is determined by the local field strength or voltage gradient E, which is in V m^{-1}. When the cross-section of the capillary and the composition of the solution in it are uniform over the length L, the field strength is the same everywhere and E can be calculated from the applied voltage as:

$$E = V_{appl} / L$$

Values of E used in practice are 20–60 kV m^{-1}.

When the voltage is applied, an electric current will flow through the capillary. Ohm's Law is approximately obeyed, i.e., the current is approximately proportional to the applied voltage. (Deviations occur due to temperature changes; see Chapter 5.) Sometimes it is more convenient to discuss the current density at a specific point in the solution. The current density j is defined as the current through a plane of unit area (j is in A m^{-2}). A variant of Ohm's Law relates the current density and the field strength at a specific point:

$$j = \kappa \cdot E \qquad (4.1)$$

where κ is the electric conductivity of the solution at that specific point. The unit of the conductivity is Ω^{-1} m^{-1} or $\mho$ m^{-1}.

Irrespective of inhomogeneities of the diameter of the capillary or the composition of the solution, the electric current must be the same at any cross-section of the capillary. When differences in conductivity exist in the capillary, as will be the case during a separation, the local field strength will vary inversely proportional with the conductivity. In Figure 4.1 this is shown schematically. This important phenomenon, which is the source of overloading in CE, will be discussed later.

Under the assumption that the diameter of the capillary and the composition of the solution are constant over the length of the capillary, the electrophoretic current can be found as:

$$i = {}^{1}/_{4}\,\pi d_c^2 \cdot \kappa \cdot V_{appl} / L \qquad (4.2)$$

When the conductivity of the BGE is known, one can calculate what current to expect. Usually the current is kept below 50 μA.

The equivalent conductivity Λ of a simple electrolyte solution has been defined as:

$$\Lambda = \frac{\kappa}{c_{eq}} \qquad (4.3)$$

Here c_{eq} is the equivalent concentration (z.c) of the ions of the electrolyte. The conductivity of a simple salt

0009-5893/00   S-20-04 $ 03.00/0

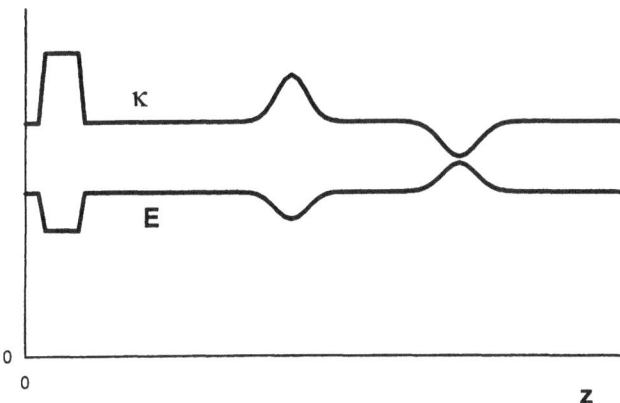

Figure 4.1
Relation between the local solution conductivity and the local field strength.

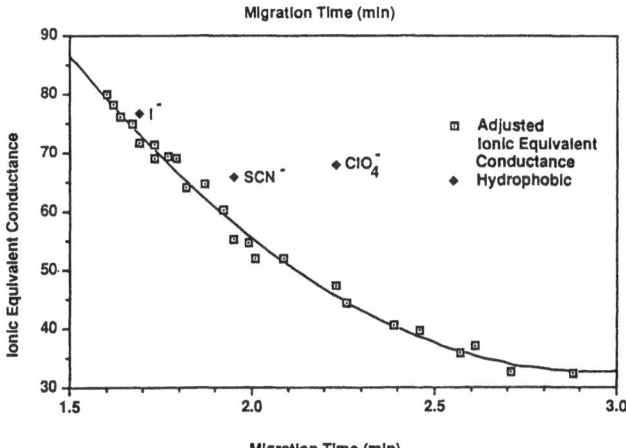

Figure 4.2
Migration times of anions plotted against ionic conductance values. Reproduced from ref. [3].

Table 4.1 Conductivities of buffer solutions that can be used as BGE in CE.

BGE composition [mol L^{-1}]	approx. pH	κ [Ω^{-1} m^{-1}]
0.01 NaCl	–	0.119
0.01 HCl	2.0	0.412
formic acid + 0.01 NaOH	3.7	0.108
acetic acid + 0.01 NaOH	4.7	0.078
MES + 0.01 NaOH	6.1	0.071
ACES + 0.01 NaOH	6.8	0.073
0.005 NaH$_2$PO$_4$ + 0.005 Na$_2$HPO$_4$	7.1	0.137
veronal + 0.01 NaOH	7.4	0.072
HEPES + 0.01 NaOH	7.5	0.067
TRIS + 0.01 HCl	8.3	0.098
boric acid + 0.01 NaOH	9.1	0.091
CAPS + 0.01 NaOH	10.4	0.088
0.01 NaOH	12.0	0.235

tabulated for a large number of ions; their mobilities (at infinite dilution) are easily calculated from these tables (see also Table 2.1). In Table 4.1 measured conductivities (at 25 °C) of solutions that could be used as BGE in CE are given. For all BGE's in the table the total salt concentration is 0.01 mol L^{-1}.

4.3 Electrolysis of the BGE in the electrode vials

Attention has to be given to the processes occurring at the electrodes. The passage of current through the solution is accompanied with electrochemical reactions at the electrodes. In general, decomposition of water takes place. The reaction scheme at the positive electrode is:

$$2\,H_2O \Rightarrow O_2(g) + 4\,H^+ + 4\,e^-$$

and at the negative electrode:

$$2\,H_2O + 2\,e^- \Rightarrow H_2(g) + 2\,OH^-$$

A vent should be present in the electrode vessels so that the evolving gasses can escape; otherwise in the long run the pressure build-up may cause the solution in the capillary to flow from one side to the other.

A more serious problem is the production of protons and hydroxide ions. The solution in the vial at the positive electrode will become more acidic; that at the negative electrode more basic. Buffering of the electrophoresis solutions is therefore always necessary. According to Faraday's Law the number of moles n of ions produced at an electrode is proportional to the amount of charge Q passed, with $n = Q/F$, where F is the Faraday constant. When for instance in an experiment a current of 10 μA is passed for 30 min, about 0.2 μmol of protons and hydroxide ions are produced in the two electrode vessels. From the equations given above it can be derived easily that in a standard CE experiment the amount of H$^+$ or OH$^-$ ions produced in the electrode vials corresponds with approximately two times the total amount of BGE salt present in the separation capillary. With a typical capillary

solution can be regarded as the sum of the contributions of the positive and negative ions:

$$\Lambda = \lambda_+ + \lambda_- \qquad (4.4)$$

The λ-contributions are called equivalent ionic conductivities. A strict relation between the equivalent conductivities and the migration times of anions in CE has been found experimentally [3], as is shown in Figure 4.2. There is a close relation between the concepts of ionic mobility and the equivalent ionic conductivity. For a specific ion μ_i quantifies the distance the ion migrates under standard conditions; λ_i the amount of electric charge-transport during migration. Therefore, the two concepts are related by:

$$\lambda_i = F \cdot \mu_i \qquad (4.5)$$

where F is the Faraday constant (96,500 Coulomb mol^{-1}). Ionic conductivities at infinite dilution have been

Table 4.2 Stability of BGE's against (positive) pH-shifts migrating upstream from the outlet vial.

Buffer type	composition [mmol L^{-1}]			μ_{eo}[a]	observation
acetate	HAc	NaAc			
	1	10		73	pH-shift
	2.5	10		66	pH-shift
	5	10		59	stable
	10	10		54	stable
acetate + NaCl	HAc	NaAc	NaCl		
	5	10	0	56	stable
	5	10	10	58	pH-shift
	5	10	20	57	pH-shift
borate	H$_3$BO$_3$	NaOH			
	10.5	10		97	pH-shift
	11	10		99	stable
	12	10		98	stable
TRIS	TRIS.HCl	TRIS			
	10	0		51[b]	stable
	10	1		80	stable
	10	10		82	stable
phosphate	NaH$_2$PO$_4$	Na$_2$HPO$_4$			
	2.5	7.5		84	pH-shift
	5	5		83	stable
	7.5	2.5		83	stable

[a]: electroosmotic mobility in 10^{-9} m^2 V^{-1} s^{-1}
[b]: with fresh BGE in inlet vial

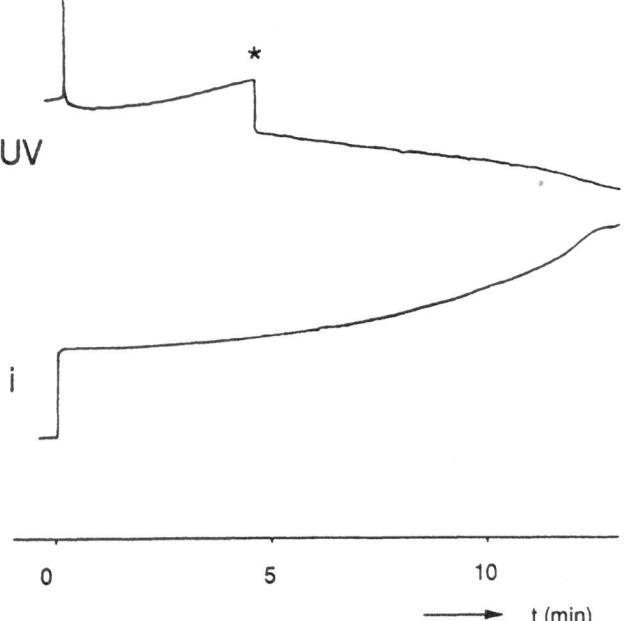

Figure 4.3
Instability of the electrophoretic current and the detector baseline by a pH-shift migrating upstream from the end vial. The asterix indicates the passage of the pH-shift through the detection window.

volume of one microliter this may not seem much, but when the volume of the electrode vessel is small or the solution insufficiently buffered, serious pH changes of the BGE in the inlet and outlet vials may result.

Since it is common practice to flush the capillary from the inlet vial before an experiment, a change of the pH of the inlet vial solution will change the effective (and apparent) mobilities of partly ionised analytes [4]. Because of the water decomposition at the electrode in this vial, the solution will have to be refreshed after a certain number of electrophoretic runs. When a 1:1 acid/base buffer is used as the BGE, and one wants to keep the effective mobilities of weakly acidic or basic analytes constant within 1%, the maximum number of runs that can be done with a single inlet vial can be estimated as:

$$\text{max.no.runs} = 5 \cdot \frac{\text{volume of capillary (in } \mu l)}{\text{volume of BGE in inlet vial (in ml)}}$$

When non-buffering salts are also present in the BGE, for instance SDS salt when MECC is applied, the pH changes in the electrode vials are even more pronounced.

With most instruments the refreshing of the inlet vial solution is easy. With some apparatus a number of positions of the autosampler tray can be reserved for vials containing the BGE; changing of the solution after a certain number of analyses is then a question of programming the run sequence. With other instruments the content of the inlet vial can be refreshed by flushing from a large storage bottle of the BGE solution.

The change of the pH of the solution by electrolysis in the outlet vial could lead to a change of the separation conditions during a run. In some cases it is possible that a shift of the pH migrates from the outlet vial, against the electroosmotic flow, into the separation capillary. This would lead to irreproducible migration times, since the magnitude of the pH shift would depend on the number of analysis already carried out with the same outlet vial solution. A migrating pH shift causes a change of the electrophoretic current during the run, because the average conductivity of the solution in the capillary changes. Moreover, such a migrating pH-shift is often visible on the detector signal as a sharp change of the baseline. An example of this alteration of the current and the accompanying detector signal are shown in Figure 4.3. Here, a shift of the pH to a higher value migrated from the end vial, against the electroosmotic flow, through a phosphate buffer used as the BGE [5].

Automatic refreshing or replacing of the outlet vial solution is not always possible with the instruments presently available on the market. However, there is in many cases also no need for this. Despite the high mobility of OH$^-$ ions (or H$_3$O$^+$-ions in case the voltage polarity is reversed), the velocity of a pH shift is generally low in a buffering BGE. As long as there is a substantial EOF in the direction of the detector, pH shifts can often not migrate against it. This was shown in experiments as reported in Table 4.2 [6]. Here, the BGE in the outlet vial, with the negative electrode, was replaced with a sodium hydroxide solution. It appeared that in many cases the system was stable, i.e., the high pH of the end vial solution could not enter the capillary.

From these and other experiments, the following conclusions could be drawn:

- the migration of a pH-shift upstream from the end vial is more likely to occur when the electroosmotic mobility is low;
- the upstream migration of pH-shifts is generally easily avoided by choosing a BGE with a proper buffer capacity;
- the presence of other, non-buffering salts in the BGE increases the possibility of pH-shifts during the run; the use of mixed buffers (e.g., phosphate-borate buffers) is therefore not recommended;
- the upstream migration of a positive pH-shift is not possible with a BGE containing (only) a buffer of the type HB^+/B, such as a TRIS buffer;
- the upstream migration of a negative pH-shift (with a reversed voltage and direction of the EOF) is not possible with a BGE containing (only) a buffer of the type HA/A^-, such as an acetate buffer;
- buffers containing multivalent ions (phosphate, citrate) are not very stable against pH-shifts; when it is not possible to refresh the outlet vial solution continually, such buffers should not be used.

References

[1] A. Guttman and N. Cooke, J. Chromatogr., 559 (1991) 285.
[2] N. Chen, L. Wang, and Y. Zhang, J. Chromatogr., 644 (1993) 175.
[3] W.R. Jones and P. Jandik, J. Chromatogr., 546 (1991) 445.
[4] H. Corstjens, H.A.H. Billiet, J. Frank, and K.C.A.M. Luyben, Electrophoresis, 17 (1996) 137.
[5] W.Th. Kok and Y. Sahin, Anal. Chem., 65 (1993) 2497.
[6] W.Th. Kok and W.J. Ozinga, publication in preparation.

5 Thermal Management

5.1 Joule heating in CE

The paper of Jorgenson and Lukacs published in 1981 [1] is often referred to as the starting point of modern CE. Still, in the years before this, several research groups [2, 3, 4] published on "free zone electrophoresis" in narrow tubes, which is essentially the same as CE. The break-through in 1981 was achieved by the improvement of the thermal management of the system, simply by decreasing the diameter of the separation channel. This scaling-down had become possible by two factors important for analytical chemistry. One factor was the development of (optically transparent) microcapillaries, first of Pyrex glass and later of fused silica. The other crucial factor for the development of CE was the availability of detectors suitable for the small volume scale employed. Jorgenson and Lukacs used on-column fluorescence detection.

Thermal problems are caused by the Joule heat generated by the electrophoretic current. Excessive heat generation can have several harmful effects.

- In wide capillaries a loss of separation efficiency can be caused by convection in the solution, generated by density differences. This effect can be counteracted by rotating the capillary around its axis [2], which however strongly complicates the instrumental set-up.

- Radial temperature differences in the solution inside the capillary cause a decrease of the separation efficiency. This effect of Joule heating on the separation efficiency has been discussed in Chapter 3. As has been shown there, under typical CE conditions the effect is usually negligible. Only with very short capillaries, using high field strengths for high-speed separations, this zone broadening contribution is of importance. Here it should be stressed that radial temperature gradients within the solution can not be opposed by a more effective thermostatting of the capillary. The cooling system of a CE instrument can only help to keep the average temperature rise in hand.

- With a rise of the average temperature of the solution inside the capillary, diffusion coefficients of the analytes are increased. This gives an increase of the zone broadening by axial diffusion, which is only partially compensated for by the increase in the speed of the analysis. As a result, plate numbers are decreased at higher temperatures [5].

- The repeatability is affected when variations occur in the efficiency of the cooling system. It has been shown that when highly conductive buffers are used, the coefficients of variation of migration times and peak areas can be decreased by, on average, a factor of two, by changing from an air-cooled to a liquid-cooled system [6, 7].

- An elevation of the average temperature can cause the decomposition or structural change of thermolabile compounds such as proteins [8, 9].

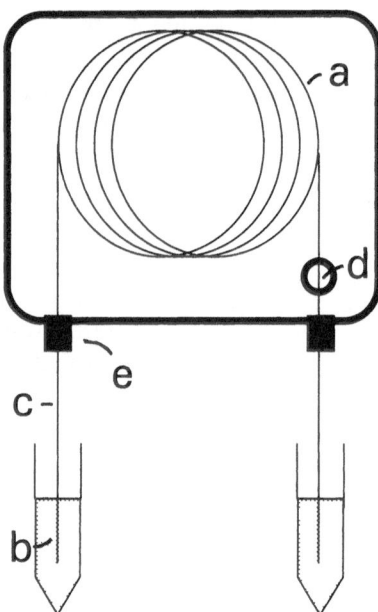

Figure 5.1
Possible cooling conditions at different parts of the capillary in a cassette-system. Cooling by (a) forced liquid; (b) stagnant liquid; (c) forced air; (d) stagnant air; (e) solid material.

- In the extreme, disruption of the electrophoretic process by air bubble formation or boiling of the solutions will occur.

As has amply been shown in practice, CE can be performed with the typical 50 μm capillaries without any cooling other than that by free convection of the surrounding air. However, for more demanding applications, in terms of separation speed, sample capacity or repeatability, the choice of the thermostatting system may be of importance. Here, some warning is appropriate on the usefulness expected of a thermostatting system. As already mentioned above, thermostatting can never decrease radial temperature differences inside the solution or over the capillary wall. Moreover, in present-day instruments it is impossible to have uniform thermostatting conditions over the complete length of the capillary. In Figure 5.1 an example is sketched of these different cooling conditions. Weak points are the parts of the capillary close to the ends and near the detector window. Some of the Joule-heat related problems itemised above will be reduced in proportion to the fraction of the capillary properly thermostatted. However, for other effects, such as the decomposition of thermolabile analytes or the eventual boiling of the solution, a 'worst-case' strategy has to be adopted. Operating conditions have to be chosen such that even at the most poorly thermostatted part of the capillary the temperature elevation is kept within limits.

5.2 Thermostatting systems

Thermal management in CE comes down to providing an effective way of dissipation for the Joule heat devel-

0009-5893/00   S-24-04 $ 03.00/0

oped. Joule heating in CE is most conveniently discussed in terms of the power generated per unit length of the capillary, P (in W.m^{-1}). This power can be expressed with a variant of Ohm's Law as:

$$P = i \cdot E = {}^1\!/_4 \, \pi d_c^2 \cdot \Lambda_b \, c_b \cdot E^2 \qquad (5.1)$$

where Λ_b is the equivalent conductance and c_b the concentration of the BGE salt. For simple buffer salts Λ_b is in the order of 10 $(\Omega \, m^{-1})/(mol \, L^{-1})$. Obviously, decreasing the capillary diameter will decrease Joule heating effects. When the Joule heating has been reduced sufficiently, a further decrease of the capillary diameter can be utilised to increase the applied field strength, thereby improving the speed and efficiency of the separation.

The problem of the heating effect occurring in free solution electrophoresis has been widely addressed from a theoretical [3, 5, 10, 11, 12, 13] and experimental point of view [6, 7, 8, 14, 15]. The average temperature elevation of the solutions during CE separations, ΔT, can be considered as proportional to the electrical power P generated per unit length of the capillary, with a proportionality constant r, the so-called heat transfer resistance. The heat transfer resistance of the solution itself (r_s), the wall of the capillary (r_w) and the surrounding thermostatting medium (r_m) are the three main contributions to r:

$$\Delta T = r \cdot P = (r_s + r_w + r_m) \cdot P \qquad (5.2)$$

The heat transfer resistance of the solution and the wall can be calculated easily. That of the solution is given by:

$$r_s = \frac{1}{4\pi\lambda_s} \approx 0.13 \qquad (5.3)$$

For the value of the thermal conductivity of the solution (λ_s), that of water is taken here. Values of r are given in K W^{-1} m. Under typical CE conditions, when the power P is usually below 1 W m^{-1}, this leads to a temperature difference between the solution in the centre of the capillary and that close to the inner wall of less than 0.1 K. The thermal resistance of the capillary wall is given by:

$$r_w = \frac{1}{2\pi\lambda_w} \ln \frac{d_{out}}{d_c} \approx 0.27 \ln \frac{d_{out}}{d_c} \qquad (5.4)$$

with λ_w as the thermal conductivity of fused silica, and d_{out} and d_c as the outer and inner diameter of the capillary, respectively. Under typical CE conditions the temperature difference over the capillary wall is less than 1 K.

The heat transfer resistance in the surrounding thermostatting medium (r_m), which is the most important contribution in practice, depends on the outer diameter of the tube and in a complicated mode on the flow regime and the properties of the cooling fluid. Even under carefully defined and controlled cooling conditions, it is not possible to derive theoretical expressions predicting the temperature elevation of the capillary. However, in

Table 5.1 Calculated and experimental heat transfer resistances of CE systems with different cooling fluids.

fluid	velocity [m s^{-1}]	r^a [K W^{-1} m]		refs.
		calculated	experimental	
still air	0	16	12–14	[5, 9]
forced air	1	4	5–7	[5, 9, 18]
still liquid	0	0.6^b	< 1	[8]
forced liquid	0.01	0.4^b	< 1	[6]

^{a} including r_b and r_w
^{b} calculated values for water

the literature empirical relations are given that relate heat transfer characteristics of heated cylinders with the experimental parameters [16, 17]. These relations have been used to estimate the heat transfer resistances of the thermostatting medium around the capillary in a typical CE set-up. The following thermostatting conditions have been compared:

- no thermostatting, i.e., cooling by the natural convection of the surrounding air;
- thermostatting by a forced air stream, with a velocity of 1 m s^{-1};
- thermostatting by a stagnant liquid;
- thermostatting by a liquid flowing with a velocity of 0.01 m s^{-1} over the capillary.

The calculated values are given in Table 5.1. The data given for liquid cooling have been calculated for water; for other liquids (Freon, toluene, ethylene glycol) similar results were obtained. Also included in Table 5.1 are the values of r found in practice, as given in (or derived from) the literature. A reasonable agreement is found between the calculated and experimental values of r. Conclusions that can be drawn from these data are as follows.

- With forced air cooling the temperature elevation can be decreased with approximately a factor of two compared to cooling by natural convection only.
- The temperature elevation with liquid cooling is approximately ten times lower than with (forced) air cooling.
- With liquid cooling the effect of forced convection, with the fluid velocities that can be applied in practice, is only small. With liquid cooling the thermal resistance of the capillary wall becomes the limiting factor anyway. Circulating the cooling liquid is mainly useful for an easier control of its (average) temperature.

When the solution in the capillary is propelled by the electroosmotic process from one cooling domain to another (e.g., from a liquid cooled part of the capillary to an air cooled part), its average temperature adapts to the new conditions within milliseconds. Therefore, it must be stressed again that often the weakest spot in the cooling system determines the overall performance, and that the maximum power that can be applied during a CE

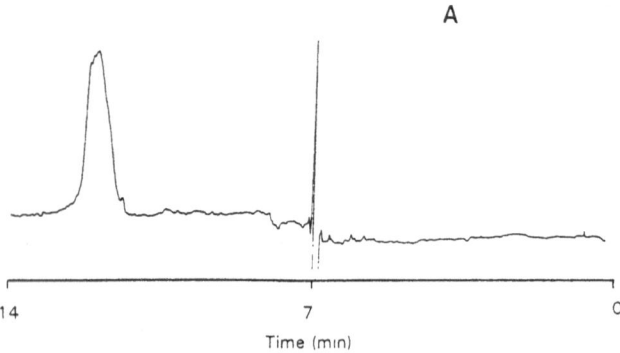

A

14   7   0

Time (min)

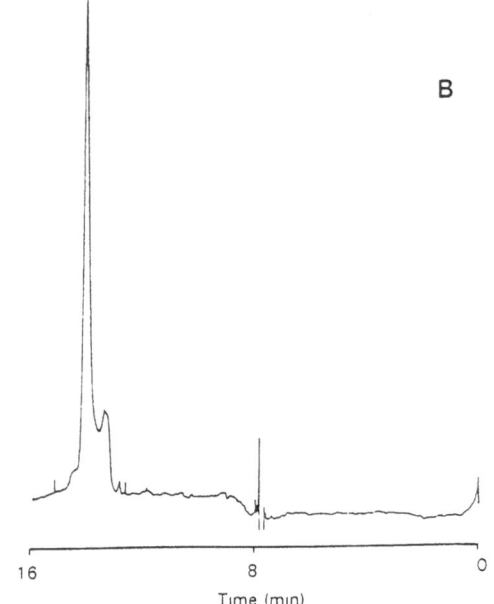

B

16   8   0

Time (min)

Figure 5.2

Influence of the thermostatting conditions on the degradation of β-lactoglobuline A during a CE separation. Cooling by (A) forced air and (B) forced helium. Reproduced from reference [9].

separation is often determined by the allowable temperature increase at a spot with only cooling by natural air convection. Interesting in this respect is that helium has a much higher thermal conductivity than air, and therefore has better cooling properties [9]. In the example shown in Figure 5.2 this is demonstrated. The decomposition of a protein during the separation is considerably less pronounced in a helium cooled system. It has even been shown that with still helium as cooling medium temperature elevations were lower than with forced air cooling. Therefore, it might be worthwhile in instrumental development to implement the option to cover "weak spots" with helium gas.

5.3 Estimation of the operating temperature

As has been discussed in the previous paragraph, the actual temperature of the solutions during a CE separation may differ considerably from ambient temperature or the set-point value of the instrument's thermostat. In some cases one would like to have an estimation of the

actual operative temperature, e.g., for troubleshooting or to be able to compare observed mobilities with literature data. Several methods have been proposed to make such an estimation [5, 9]. These methods are all based on the comparison of the value of an experimental parameter (current, mobility, plate number) observed under the operative conditions with that observed at low field strength. For the latter case it is assumed that the temperature is at the ambient or set-point value. A problem with these methods is that it is often difficult to determine at what field strength the temperature elevation is still negligible. In the following an approach will be presented in which this problem is avoided. Another advantage of this approach is that it gives directly the value of the total resistance against heat transfer of the system r, which relates the temperature elevation with the electric power during the separation.

It has been generally observed that, to a good approximation, the equivalent conductance of aqueous salt solutions increases linear with the temperature in the range from 20 to 60 °C:

$$\Lambda = \Lambda_0 \left(1 + \alpha_{th} \cdot \Delta T\right) \quad (5.5)$$

where Λ_0 is the conductance at some reference temperature (e.g., 20 or 25 °C) and ΔT the increase of the temperature above the reference value. For all simple salt solutions the value of α_{th} is close to 0.02; this same value has also been found for typical CE buffer solutions [18]. By combining Equation (5.5) with (5.1) and (5.2) the following expression can be obtained for the relation between the electrophoretic current and the applied field strength:

$$\frac{i}{^1/_4\,\pi d_c^2 \cdot E} = \Lambda_0\, c_b \cdot \left(1 + \alpha_{th} \cdot r \cdot i \cdot E\right) \quad (5.6)$$

The thermal characteristics of the instrumental set-up can now be evaluated as follows.

- Fill the capillary with an electrolyte solution. The salt concentration or conductivity of this solution should be relatively high, at least as high as that of the BGE to be used in the application to be run.

- Measure the electrophoretic current with a number of values of the applied voltage (e.g., 5, 10, 15, 20, 25, and 30 kV).

- For each applied voltage, calculate the power per meter P as $(i \cdot V_{app})/L$ and the solution conductivity κ as $(i \cdot L)/(^1/_4\pi d_c^2\, V_{app})$. When κ is plotted against P, a straight line should be obtained.

- Find the intercept and the slope of the (straight) line of κ vs. P by linear regression.

- The intercept gives the solution conductivity κ at the ambient temperature (in $\Omega^{-1}\,m^{-1}$).

- The thermal resistance of the set-up is now found as: $r = 50 \cdot$ slope / intercept

- The value for r obtained gives an indication of the overall quality of the temperature control. It can be

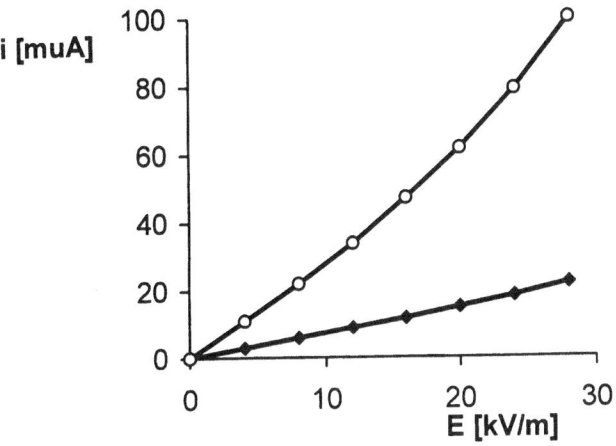

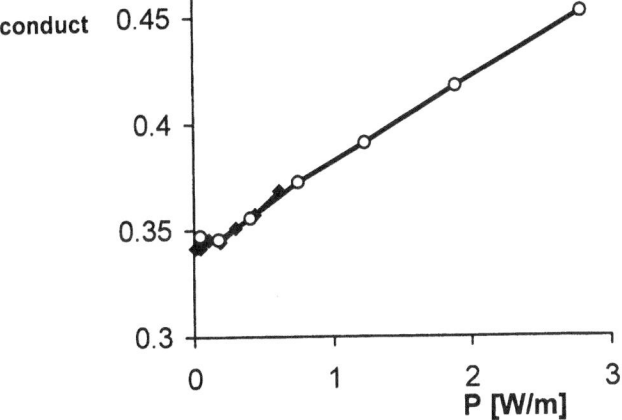

Figure 5.3
Interpretation of the current-voltage relation in CE for the calculation of the average heat transfer resistance r and the operating temperature. (A) Observed current as a function of the applied voltage. (B) Calculated solution conductivity as a function of the applied power. Data from ref. [19]. BGE: 0.03 mol L^{-1} NaCl; capillary diameter: ($\blacklozenge$) 50 and ($\bigcirc$) 100 μm.

An example of this approach is given in Figure 5.3. Values of r obtained here are 6.5 and 5.9 for 50 and 100 μm capillaries, respectively. The calculated temperature increase at 30 kV applied voltage is 4 K and 20 K, respectively.

In principle, these experiments have to be carried out only once for a specific set-up. A change of the buffer composition or concentration, or of the applied field strength, will not affect the heat dissipation characteristics. To estimate the temperature elevation under the new conditions only the value of P has to be calculated from the observed current. It must be kept in mind, however, that the estimated value of ΔT is an average over the complete length of the capillary. Locally, higher temperature elevations may exist.

References

[1] J.W. Jorgenson, and K.D. Lukacs, Anal. Chem., 53 (1981) 1298.
[2] S. Hjerten, Chromatogr. Rev., 9 (1967) 122.
[3] R. Virtanen, Acta Polytech. Scand., 123 (1974) 1.
[4] F.E.P. Mikkers, F.M. Everaerts, Th. P.E.M. Verheggen, J. Chromatogr., 169 (1979) 11.
[5] J.A. Knox and K.A. McCormack, Chromatographia, 34 (1994) 207.
[6] Y. Kurosu, K. Hibi, T. Sasaki, and M. Saito, J. High Resolut. Chromatogr./Chromatogr. Commun., 14 (1991) 200.
[7] J.P. Landers, R.P. Oda, B. Madden, T.P. Sismelich, and T.C. Spelsberg, J. High Resolut. Chromatogr./Chromatogr. Commun., 15 (1992) 517.
[8] R.J. Nelson, A. Paulus, A.S. Cohen, A. Guttman, and B.L. Karger, J. Chromatogr., 480 (1989) 111.
[9] A. Cifuentes, W.Th. Kok, and H. Poppe, J. Microcolumn Sep., 7 (1995) 365.
[10] J.H. Knox, and I.H. Grant, Chromatographia, 24 (1987) 135.
[11] W.Th. Kok, Chromatographia, 24 (1987) 442.
[12] E. Grushka, R.M. McCormick and J.J. Kirkland, Anal. Chem., 61 (1989) 241.
[13] J.A. Knox and K.A. McCormack, Chromatographia, 34 (1994) 215.
[14] G.J.M. Bruin, P.P.H. Tock, J.C. Kraak and H. Poppe, J. Chromatogr., 517 (1990) 557.
[15] A. Vinther and H. Soeberg, J. Chromatogr., 559 (1991) 27.
[16] W. H. McAdams, Heat Transmission, Third Edition, (McGraw-Hill, New York, 1954).
[17] S.T. Hsu, Engineering Heat Transfer, (D. Van Nostrand Company, Inc., New Jersey, 1963).
[18] M.S. Bello, M. Chiari, M. Nesi, P.G. Righetti, and M. Saracchi, J. Chromatogr., 625 (1992) 323.
[19] A. Cifuentes, X. Xu, W.Th. Kok, and H. Poppe, J. Chromatogr.A, 716 (1995) 141.

used to estimate the temperature elevation for all applications with the set-up in use, as long as the capillary outer diameter is the same. For a specific application, the operating temperature can be estimated as:

$$T = T_{amb} + r \cdot (i \cdot V_{app}/L)$$

where T_{amb} is the ambient or set-point temperature.

6. Capillaries and the Electroosmotic Flow

6.1 The origin of the electroosmotic flow

The velocity of the electroosmotic flow (EOF) is a crucial parameter for separations by capillary electrophoresis. It determines the range of analyte mobilities that can be handled in a single run as well as the run time. Moreover, variations in the EOF, between different capillaries and between different runs on the same capillary, are the main source of the irrepeatability and irreproducibility of the elution times so often encountered in CE. The electroosmotic mobility is in principle governed by the chemical condition of the inner surface of the capillary and by the composition of the background electrolyte (BGE) used for the separation. Still, even with careful control of these parameters, large variations in the EOF are found in practice.

The origin of the electroosmotic process is in the ionization of the inner surface of the capillary wall. When silica is contact with an aqueous solution (the BGE) with a pH higher than 2–3, silanol groups on the surface are partly ionized:

$$-SiOH(s) \leftrightarrow -SiO^-(s) + H^+(aq)$$

To describe this dissociation several values for the pK_a of the silanol groups have been given in the literature. However, the dependency of the EOF on pH does not resemble a simple titration curve with a single pK_a-value (see also next paragraph). Rather, in the pH-range from 2 to 9 a gradual increase of the EOF, and therefore of the surface ionization, is seen. This may indicate that there is a strong heterogeneity in the acidity of silanol groups.

Following the electroneutrality principle, the negative charge on the capillary inner surface is neutralized by an excess of positive ions from the BGE close to the surface, preferentially attracted by electrostatic forces. The negative and positive charges form an electrical double layer on the surface. According to the Stern model for the electric double layer [1], the double layer is divided in two parts: an inner, compact layer and an outer, diffuse layer. The compact layer comprises specifically adsorbed and the first layer of non-specifically adsorbed ions. It has been shown that the capacity of this inner layer is relatively constant [2], which means that the surface charge density in the compact layer is an approximately constant fraction of the charge density on the solid surface. In the next, for simplicity the compact layer charge is therefore assumed to be comprised in the surface charge of the capillary wall (σ_w). The charge in the diffuse layer (σ_d) is equal to (but of opposite sign) the sum of these two charges.

The negative charge on the silica surface gives it a negative potential. When the BGE is composed of a single 1:1 salt with a concentration c_{BGE}, the surface potential ϕ_0 (taken here on the interface between the compact layer and the diffuse layer) is related to the surface charge σ_w by:

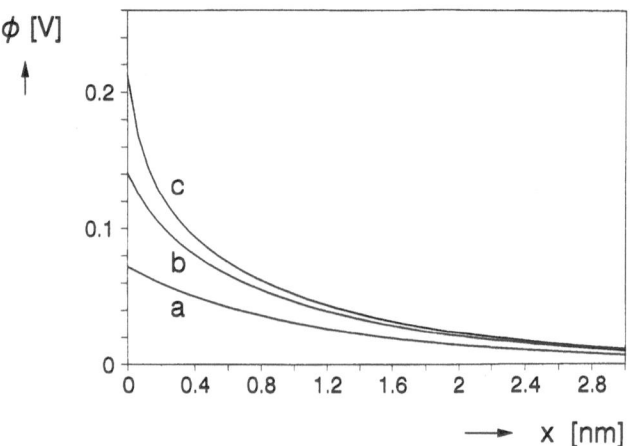

Figure 6.1

Course of the potential in the diffuse layer as a function of the surface charge density; $\sigma_d = 0.05$ (a), 0.2 (b) and 0.8 (c) C m^{-2}. Ionic strength is 0.05 mol L^{-1}. Reproduced from ref. [8].

$$\phi_0 = \frac{2RT}{F} \sinh^{-1} \left(\frac{\sigma_w}{(8RT\varepsilon_{aq}\varepsilon_0 c_{BGE})^{1/2}} \right) \tag{6.1}$$

where ε_{aq} is the permittivity of the solution; the other symbols have their usual meaning. Only for low surface charges and potentials (< 25 mV) the surface potential is proportional to the surface charge; for higher values of σ_w the relation of ϕ_0 with it increases less steep. Equation (6.1) shows that the surface potential decreases with increasing salt concentration of the BGE.

In the diffuse layer the solution potential falls off, due to the presence of the excess of positive charges, approaching zero in the bulk of the solution. The slope of the potential *vs.* distance curve depends on the value of ϕ_0: with higher surface potentials the potential drops more quickly. In Figure 6.1 the course of the potential in the diffuse layer is shown as a function of the surface charge density. Moreover, the ionic strength of the solution has a large influence; with increasing salt concentration the potential drop in the diffuse layer becomes steeper. The diffuse layer thickness is approximately inversely proportional to the square root of the ionic strength.

In the discussion of the electroosmotic flow in fused silica capillaries it is generally assumed that there is a stagnant layer of solution with a finite thickness δ on the inner wall of the capillary. The electrostatic force of the applied axial electric field on the excess charge in the solution outside this stagnant layer causes the EOF. The differentiation between a stagnant and a mobile layer is another one than that between the compact and the diffuse layer. In Figure 6.2 this difference is illustrated. The existence of a stagnant layer of (aqueous) solution on a charged surface is usually explained as the result of strong dipole interactions between water molecules in the high electric fields close to the surface, causing a locally increased viscosity of the solution. However, electrokinetic phenomena on silica, the standard capil-

Chromatographia Supplement, Vol. 51, 2000   Capillaries and the Electroosmotic Flow

0009-5893/00   S-28-08 $ 03.00/0

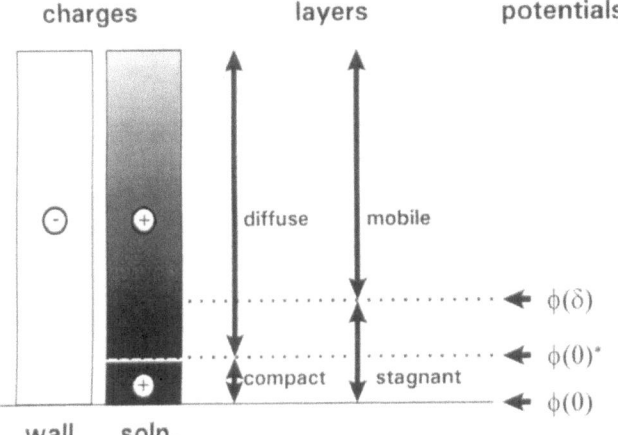

Figure 6.2

The electrical double layer and electrokinetic phenomena. $\phi(0)$ is the real surface potential, $\phi(0)^*$ that on the surface of the compact layer.

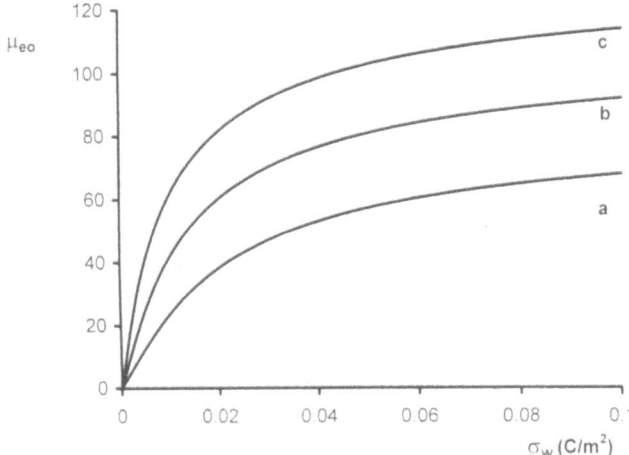

Figure 6.3

Dependency of the electroosmotic mobility on the surface charge of fused silica. Ionic strength 10 (a), 3 (b) and 1 (c) mmol L^{-1}.

lary material in CE, have been found to be unusually small in comparison to other solids with a similar surface charge density [3]. As a possible explanation for this, the presence of a gel-like structure on top of the silica surface has been proposed, extending the thickness of the stagnant layer [4, 5]. Other models developed to reconcile the high surface charge of silica with its low zeta potential account for the complexation of 'indifferent' cations with the anionic groups on the surface [6, 7] (see paragraph 6.2).

It can easily be shown that the electroosmotic mobility is independent of the exact potential path in the mobile solution; μ_{eo} is related to the solution potential at the outside of the stagnant layer $\phi(\delta)$, also called the zeta potential, by:

$$\mu_{eo} = -\frac{\varepsilon_{aq}\varepsilon_0}{\eta} \cdot \phi(\delta) \qquad (6.2)$$

where η is the viscosity of the solution. For aqueous solutions at 25 °C the relation can be written as:

$$\mu_{eo} = -7.8 \times 10^{-7} \cdot \phi(\delta)$$

with μ_{eo} in m^2 V^{-1} s^{-1} and $\phi(\delta)$ in V. Unfortunately, there is no way to measure the zeta potential or the thickness of the stagnant layer directly. Fitting empirical data on electroosmotic velocities with theoretical models gives for the stagnant layer thickness δ a value in the order of 0.8 nm [8]. With this value for δ, Figure 6.3 has been constructed, showing the predicted relation between the surface charge σ_w and the electroosmotic mobility. From this figure it can be concluded that:

- only for very low surface charges the electroosmotic mobility increases linearly with σ_w;
- with higher surface charges, μ_{eo} approaches asymptotically to some maximum value; this virtual maximum is reached at values of σ_w well below the maximum of surface ionization that can be reached on silica (which is in the order of 0.8 C m^{-2});
- the ionic strength of the solution has a strong effect on μ_{eo}.

Several explanations can now be given for the strong influence of the origin, the pretreatment and the history of fused silica capillaries as has been observed in practice, especially at low pH. The manufacturing process of the capillaries or pretreatment procedures (etching) may have an effect on the microstructure of the inner wall [9], thereby influencing the orientation and dissociation properties of the silanol groups. Impurities (metal ions) fixed in the silica matrix may effect specific adsorption of counter-ions, and thereby the compact layer charge. Etching procedures may effect the possible presence of gel-like structures on the surface, and with this the stagnant layer thickness [10]. Specifically adsorbed matrix components of previous samples may have changed the surface charge of the silica. It will be obvious that a careful control of these effects is crucial to obtain repeatable and reproducible peak times in CE.

6.2 Solution effects on the EOF

As discussed in the previous paragraph, there are three main parameters that determine the electroosmotic mobility: the surface charge of the inner capillary wall (including the compact layer), and the ionic strength and viscosity of the solution. All parameters depend on the composition of the BGE.

Influence of the pH

With a BGE pH below 3, the electroosmotic mobility is usually very low. It has even been reported that the EOF at very low pH is in the direction of the positive electrode. From pH 3 on the electroosmotic velocity starts to increase with the pH, reaching a maximum at pH 7–8. The slope of the curve is lower than for a typical pH

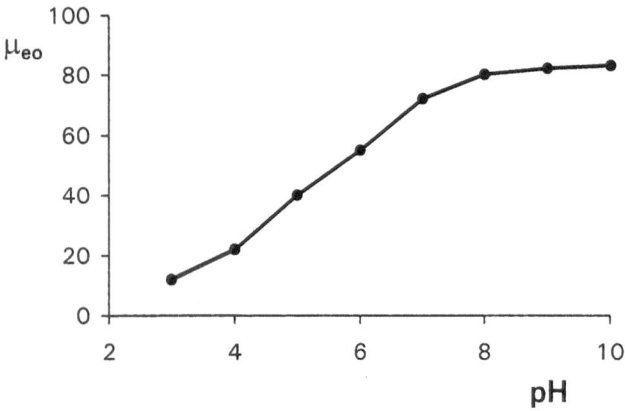

Figure 6.4
Influence of the pH of the BGE on the electroosmotic mobility in a fused silica capillary. Ionic strength 10 mmol L^{-1}. Data have been corrected for temperature effects.

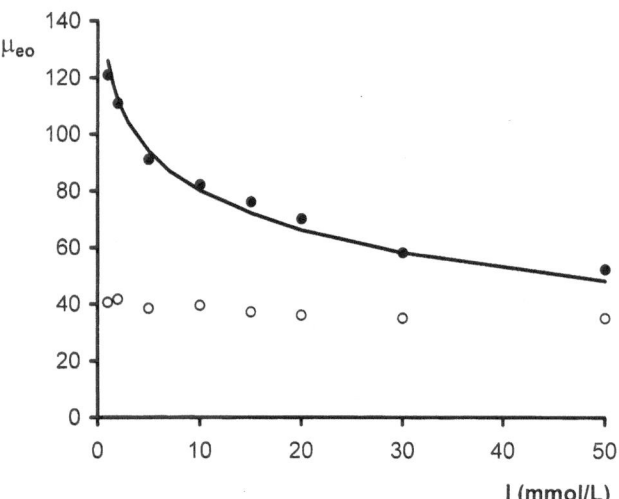

Figure 6.5
Influence of the ionic strength of the BGE on the electroosmotic mobility in a fused silica capillary at pH 8.6. Data have been corrected for temperature effects. Solid dots: experimental data on μ_{eo}; solid line: theoretical expectation; open dots: ionic mobility of 1-hydroxy-4-naphthalene sulfonic acid.

sigmoid curve; this reflects the heterogeneity of the silica groups on the surface. A typical μ_{eo} *vs.* pH curve is shown in Figure 6.4. The exact values that are measured depend on the design of the experiment. The EOF shows hysteresis, i.e., its velocity depends on the order of measuring with different buffers [11].

Stable and reproducible electroosmotic velocities are more easily obtained at high pH than at low pH values. This may be related to the high surface charge of silica at high pH; at high surface charges μ_{eo} becomes independent of σ_w (see Figure 6.2). Another explanation is that with a high pH BGE the inner surface of the capillary is continuously etched; specific adsorption of BGE impurities or sample matrix components is thereby opposed.

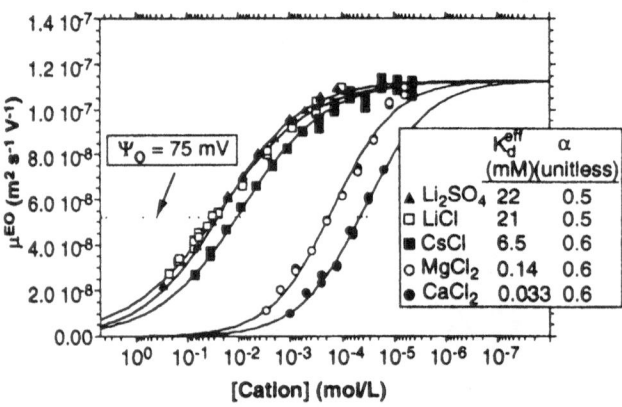

Figure 6.6
Influence of the type and concentration of the BGE cation on the electroosmotic mobility. Reproduced from ref. [14].

Influence of the ionic strength

The influence of the ionic strength on the EOF is twofold: (i) a high ionic strength decreases the surface potential of the silica surface, and (ii) with a high ionic strength a larger part of the diffuse double layer charge falls within the stagnant layer. Figure 6.5 shows experimental data on the electroosmotic mobility measured with a high-pH BGE. For comparison, data are included on the ionic mobility of a strong single-charged anion. A high ionic strength also decreases the ionic mobility; however, this effect is much smaller than for μ_{eo}, and only significant at high salt concentrations.

Influence of the BGE cation type

In most studies on this subject it has been found that with simple 1:1 salts the zeta potential and electroosmotic mobility are virtually independent of the type of cation in the BGE. Apparently, simple single-charged cations are (almost) completely precluded from specific adsorption within the compact layer. The non-specific effect of the cation concentration can also be described as an ion-exchange process with the protons of the silanol groups on the surface, with a Nernstian contribution to the zeta potential [12]. Divalent cations on the other hand, give a much stronger depression of the EOF than is expected on basis of their influence on the ionic strength of the BGE [13, 14]. This can be explained as the result of specific ion-exchange interactions with the silanol groups in the compact layer. The resulting electroosmosis can be modeled using empirical dissociation constants for the silanoxy-cation complexes. As can be seen in Figure 6.6, the binding of, e.g., magnesium and calcium ions to the silanoxy groups is over two orders of magnitude higher than that of alkali metal ions. Therefore, a reduction of μ_{eo} is already visible at sub-millimolar concentrations of these ions in the BGE.

Influence of organic modifiers

The addition of an organic modifier to the BGE generally decreases the electroosmotic mobility. In an ex-

Capillaries and the Electroosmotic Flow

tensive study on the effect of various organic solvents on the electroosmotic mobility [15] is was shown that this EOF reduction is the result of different, often synergetic, factors.

- In mixtures of water and an organic solvent the (average) dissociation constant of the silanol groups is lower than in a purely aqueous solution. This means that for a specific buffer pH, usually measured in the aqueous solution before mixing with the organic solvent, the surface ionization decreases with increasing organic modifier content of the BGE. The inflection point of the μ_{eo} vs. pH plot shifts from approximately pH 5.3 in purely aqueous solutions to approximately 6.5 and 7.5 for solutions containing 50 % (v/v) acetonitrile or methanol, respectively.

- Adding an organic solvent to an aqueous solution decreases the permittivity (dielectric constant) of the solution. This has a reducing effect on μ_{eo} (see Equation 6.2).

- The electroosmotic mobility is inversely proportional to the solution viscosity. Mixtures of water and an organic solvent often have a higher viscosity than water or the organic solvent alone. This effect is especially pronounced with water-alcohol mixtures. A 50 % (v/v) methanol mixture has a viscosity almost twice that of water. Exceptions are acetonitrile and acetone; with these solvents the viscosity decreases steadily when going from 100 % water to 100 % organic solvent.

The three effects together have an influence on the electroosmotic velocity as shown in Figure 6.7. The data given in this figure have been obtained at high pH (9–12), and can be regarded as the maximum values that can be obtained with the various solvent mixtures at an ionic strength of approximately 20 mmol L^{-1}.

6.3 Dynamic modification of the capillary

There can be one of several reasons to look for methods to alter or control the EOF in a CE capillary:

- a high electroosmotic mobility can be desirable for fast analyses;
- a low EOF can be favorable for the resolution of closely spaced peaks (see paragraph 3.2);
- a high EOF can be required for the simultaneous separation of cationic and anionic analytes;
- for mixtures of slow and very fast anions (inorganic anions) it may be necessary to reverse the direction of the EOF towards the positive electrode [16];
- in many cases it may be desirable to get a more reproducible electroosmotic velocity, irrespective of its value.

The most obvious way to change the EOF is by altering the pH, ionic strength or organic modifier content of the BGE. However, the requirements on the BGE in respect to the EOF are not always compatible with those imposed to obtain a satisfactory separation of the analytes.

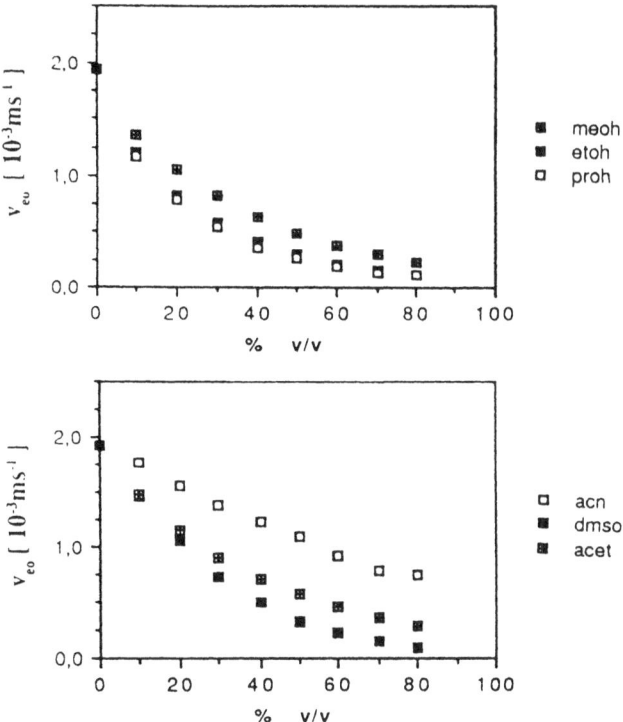

Figure 6.7
Variation of the electroosmotic velocity with solvent composition. Solvents: methanol (meoh), ethanol (etoh), 2-propanol (proh), acetonitrile (acn), dimethyl sulfoxide (dmso) and acetone (acet). Reproduced from ref. [15].

Moreover, changing the EOF through the BGE composition is easier than making it more reproducible.

The most popular method of dynamic modification of the inner wall of fused silica capillaries is by adding a surfactant to the BGE. With a cationic surfactant such as cetyltrimethylammonium bromide (CTAB) the EOF can be suppressed and even reversed [17, 18]. Two models have been proposed to describe the influence of a cationic surfactant on the EOF. Both models assume the formation of a bilayer or hemimicelle layer of surfactant molecules on the silica surface, changing its charge from negative to positive. In the first model it is assumed that at low surfactant concentration in the BGE first a single layer of adsorbed cations is formed (see Figure 6.8A). This layer is bound to the surface by electrostatic interaction with the negative silanoxy groups. With increasing modifier concentration a second layer of cationic surfactants is deposited on the first layer by hydrophobic interaction. The second model (Figure 6.8B) assumes the formation of so-called admicelles (surfactant pairs) in the solution, that are deposited on the negatively charged silica surface. An extensive study indicated that it depends on the ionic strength and pH of the BGE which model is the most adequate to describe the experimental findings [19]. The following general rules describe the effect of cationic surfactant modifiers on the EOF:

- with increasing modifier concentration the EOF is first suppressed and finally reversed, reaching a con-

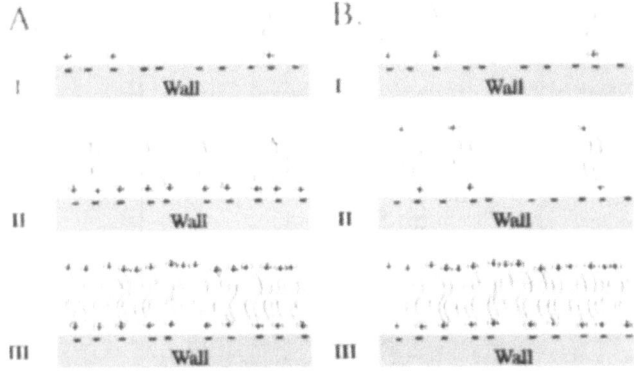

Figure 6.8

Models for the formation of bilayers of surfactants on the capillary inner surface. For explanation see text. Reproduced from ref. [19].

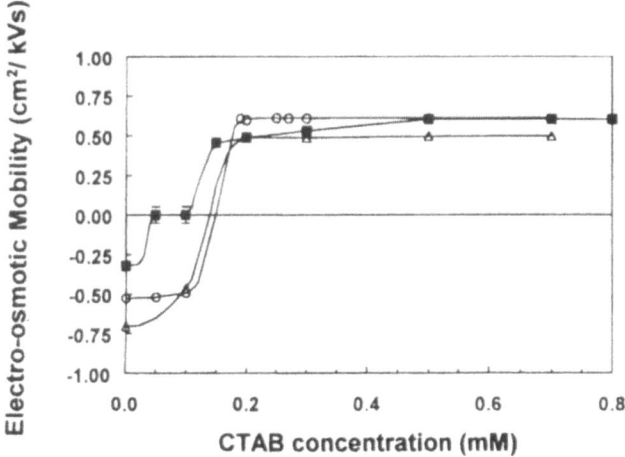

Figure 6.9

Effect of the CTAB concentration on the electroosmotic mobility at pH 3.5 (■), 7.2 (○) and 9.0 (△). Reproduced from ref. [19].

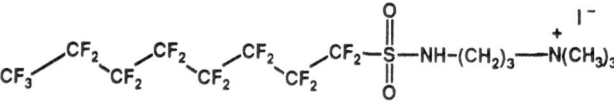

Figure 6.10

Structure of the cationic fluorosurfactant FC 134. Reproduced from ref. [20].

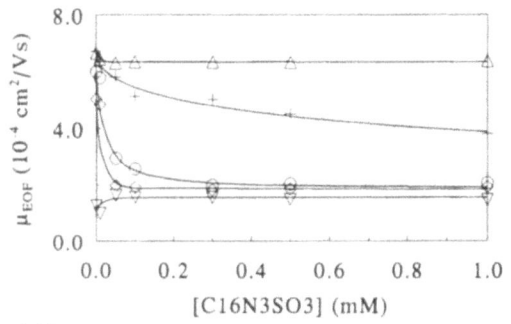

Figure 6.11

Effect of the concentration of a zwitterionic surfactant (C16N3SO3) on the electroosmotic mobility at pH (▽) 3.0, (◇) 6.0, (○) 8.0, (+) 10.0 and (△) 11.0.. Reproduced from ref. [21].

stant value above some critical modifier concentration (see Figure 6.9);

- this critical modifier concentration is strongly related to the critical micelle concentration (cmc) of the surfactant;
- the maximum value of the (reversed) electroosmotic mobility is almost independent of the pH of the BGE, for pH values in the range 3–11;
- the (maximum) electroosmotic mobility is only slightly influenced by the ionic strength of the BGE; the ionic strength effect on the EOF is much smaller than with unmodified fused silica capillaries;
- the magnitude of the resulting reversed EOF depends on the type of counter-anion used in the BGE; with organic anions or sulfate ions a lower μ_{eo} is obtained than with singly-charged inorganic counter-anions.

One of the most effective cationic surfactants used for flow reversal is Fluorad FC 134, a fluorosurfactant whose structure is depicted in Figure 6.10 [20]. It has a very low cmc and shows a strong interaction with silanoxy groups, so that it can be applied in low concentrations in the BGE (in the order of 10 ppm). This has the advantage that its presence in the separation system is not likely to cause denaturation of proteins.

When one wants to decrease or stabilize the EOF, without reversing it, a zwitterionic surfactant can be used as BGE modifier [21]. Figure 6.11 shows the effect of the concentration of such a surfactant on the electroosmotic mobility. With zwitterionic modifiers some of the same observations have been made as with cationic surfactants. A constant μ_{eo} is obtained above a certain modifier concentration, which is related to the cmc of the surfactant, and the EOF is independent of the pH in a rather broad range.

Anionic surfactants generally do not affect the EOF in bare fused silica capillaries. Apparently, their interaction with the (negatively charged) silica is negligible. However, when the capillary wall is first covalently modified with a hydrophobic surface layer, an anionic surfactant can also be used to stabilize the EOF [22]. Their adsorption to the (modified) wall increases the EOF.

6.4 Coated capillaries

In the past years a considerable research effort has been put in the development and evaluation of capillaries with a permanently modified inner surface. Most of this work was not inspired by an intention to change the EOF, although this is usually a result of the surface modification, but more by the wish to decrease the adsorption of analytes, and especially proteins, on the capillary wall. Even a weak adsorption of a high-molecular weight analyte leads to a strongly reduced separation efficiency. Proteins often adsorb on fused silica by a combined effect of electrostatic interaction, hydrogen bonding and hydrophobic interaction. By manipulation of the BGE composition (using a high pH, a high ionic strength, a surfactant or an organic modifier) it is possible to oppose some of these interactions. Often, however, the remaining adsorption effects still result in broad, tailing peaks except for a few selected model proteins.

Several approaches have been followed to prepare a permanent layer on the surface of the capillary to prevent its interaction with the analytes, with strongly varying success.

Covalent attachment of functional groups

The silane chemistry that had been developed for the modification of HPLC stationary phases has also been employed to functionalize the inner surface of CE capillaries. With C_8 or C_{18} groups bound to the surface the EOF is strongly reduced compared to bare silica, but still an influence of the pH is observed [23]. Hydrophilic groups (epoxydiol, maltose) have been attached to the silica in an attempt to reduce protein adsorption [24]. The main problem with these and comparable modifications is the limited pH range that can be used and their limited stability. At higher pH the silica bonds are readily hydrolyzed. Moreover, often the silica surface is not completely shielded, so that adsorption of analytes is not completely prevented.

Modification with polymer layers

The covalent binding of polymers to the silica surface, anchored by a bifunctional silane reagent, provides better shielding of the silica surface. In some cases prepolymerized chains, such as polyethylene glycol (PEG) or dextrans [25], have been attached to the surface, but in most cases the polymerization is performed *in situ*. Polyacrylamide coatings are relatively easily to prepare. A typical procedure involves [11]:

- vinyl modification of the silica, by filling the capillary with trichlorovinylsilane and heating at 120 °C for 18 h;
- rinsing and drying;
- flushing the capillary with a acrylamide solution in dichloromethane, also containing a suitable cross-linker, initiator and/or catalyst;
- applying a slight vacuum to suck out the excess of acrylamide solution, leaving a thin film on the surface;
- polymerization by heating at 120 °C for 3 h;
- washing and conditioning.

By the choice of the monomer, or copolymerization of different monomers, the properties of the polymer layer (linear/crosslinked; anionic/cationic/neutral; hydrophilic/hydrophobic) can be controlled. Figure 6.12 illustrates the improvement of the reproducibility of the EOF that can be obtained by using a polyacrylamide coated capillary. The type of anchoring group used is very important for the quality and stability of the polymer coating [26].

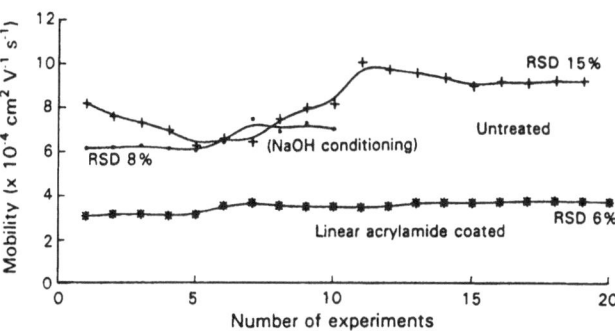

Figure 6.12
Reproducibility of the EOF with uncoated capillaries, with and without etching between runs, and with polyacrylamide coated capillaries. . Reproduced from ref. [11].

Polyacrylamide deposited on the silica surface without covalent attachment is not stable. However, it is not always necessary to bind a polymer covalently to the silica surface. Polyvinyl alcohol (PVA) for instance can be immobilized on a bare silica surface just by heat treatment [27]. PVA is normally water-soluble, so that it can be flushed into the capillary as an aqueous solution. After removal of the excess solvent the layer adsorbed on the inner surface of the capillary can be made insoluble by heating it to 140 °C for some hours.

Chemical characterization of the polymer layers in CE capillaries is difficult because of the small surface area and amount of material involved. The quality of different procedures and capillaries have to be evaluated purely empirically, e.g., by studying the EOF or the plate numbers obtained when separating proteins.

6.5 Alternative capillary materials

The problems with the irreproducibility of the EOF in fused silica capillaries have inspired a number of researchers to look for other capillary materials, mostly polymeric hollow fibers. Polypropylene has been used in a number of studies; the advantage of this material is that it is transparent for UV light. When non-transparent fibers are used, a small piece of fused silica can be inserted between two pieces of the fiber, by means of PTFE sleeves, which can then be used as detection window.

A study on the EOF in fibers made of a variety of polymer types [28] showed, surprisingly, that the electroosmotic mobility is of the same order of magnitude as in fused silica (see Figure 6.13). The surface charge of organic polymeric materials must be considerably lower than that of silica at higher pH values. Still, some charge may result from the ionization of functional groups on the surface or from the adsorption of buffer ions. As was shown in Figure 6.3, a relatively low surface charge is sufficient to cause a considerable EOF. A further explanation for the high EOF found in polymeric capillaries can be the absence of a stagnant water layer, as is assumed to exist on silica surfaces.

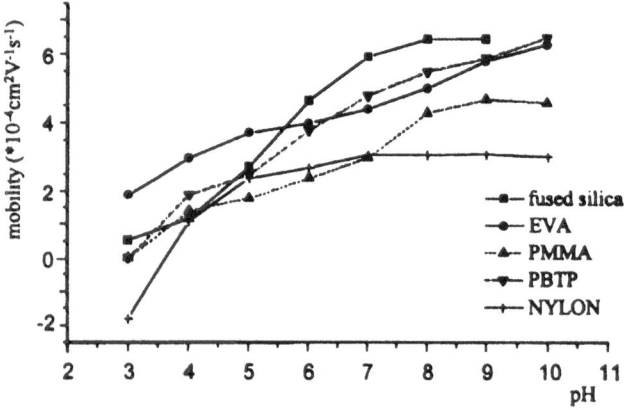

Figure 6.13
Electroosmotic mobilities in polymer capillaries as a function of the BGE pH. Reproduced from ref. [28].

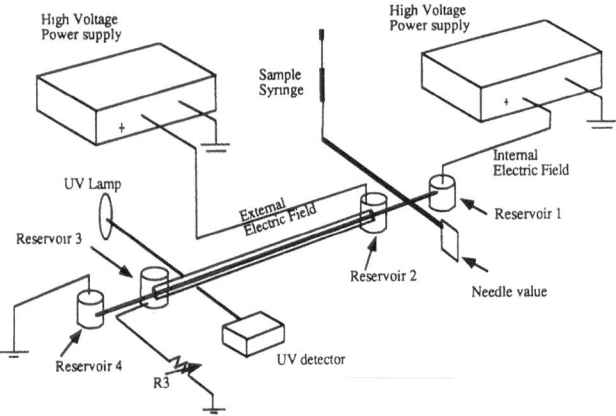

Figure 6.14
Schematic view of the set-up used to apply an external radial field which is constant over the length of the CE capillary. Reproduced from ref. [33].

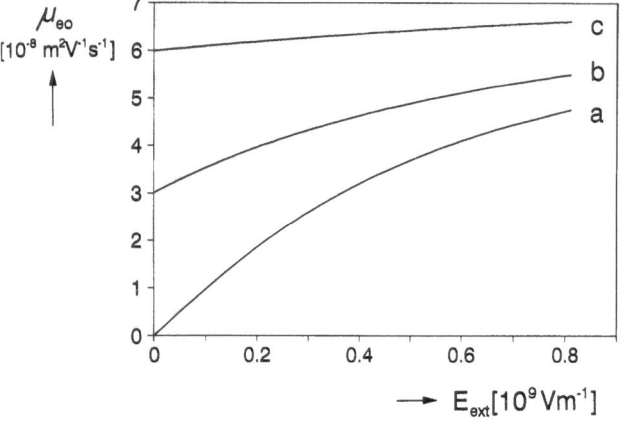

Figure 6.15
Calculated influence of an external radial field on the electroosmotic mobility. Ionic strength 10 mmol L^{-1}; $\mu_{eo^0} = 0$ (a), 30 (b) and 60 (c) $\times 10^{-9}$ m^2 V^{-1} s^{-1}. Reproduced from ref. [8].

Fused silica is still by far the most popular capillary material, due to its excellent optical properties, the low costs and its availability in a wide selection of diameters. Still, a number of advantages can be mentioned for polymeric materials. The electroosmotic mobility in polymeric fibers is less dependent on the solution pH than in fused silica capillaries. Moreover, with polymeric materials it is easier to manipulate the EOF by dynamic coating of the wall. A better control of the adsorption of surfactants is possible, e.g., based on hydrophobic interaction. Moreover, the polymeric surfaces can also be modified with anionic surfactants, giving a high EOF over a wide pH range [29].

6.6 External control of the EOF

In 1990 Lee and coworkers [30] found a way to change the EOF, and even to reverse its direction, by applying an external radial electric field. Later, external EOF control attracted the attention of a number of research groups (e.g., [31, 32]). Basically, the technique of external EOF control consists of the application of a high voltage radially over the wall of the capillary, in addition to the longitudinal field for the separation. The radial field is applied by means of a second high-voltage source between the solution inside the capillary and its outer surface. For the connection with the outer surface conductive paint or a concentric tube filled with an electrolyte solution have been used. In later versions provisions have been made to ensure that the voltage drop over the wall of the capillary was constant over (a part of) its length, by having an appropriate current flowing through the outside conductor [33]. Figure 6.14 shows such a set-up schematically. A radial voltage of several kV appeared to influence the EOF significantly when a BGE with low pH was used. With a high pH-BGE, when the electroosmotic mobility was already high, the radial field was generally found to have much less influence.

To explain the influence of the external radial field a theoretical model has been formulated, based on electrostatic theory [8]. It can be shown that an electric field applied perpendicular to the charged surface of the capillary wall increases the charge of the diffuse layer:

$$\sigma_d = -\sigma_w + \varepsilon_{si}\varepsilon_0 E_{ext} \tag{6.3}$$

where ε_{Si} is the permittivity of silica and E_{ext} the external field strength at the surface. The charge density in the diffuse layer is in this model composed of two independent contributions: one related to the chemical processes on the wall material (σ_w), and one representing the charge separation caused by the external radial field. In the cylindrical geometry of the capillary, E_{ext} can be calculated from the applied radial voltage V_{ext} as:

$$E_{ext} = \frac{V_{ext}}{R_{in} \ln (R_{out}/R_{in})} \tag{6.4}$$

Capillaries and the Electroosmotic Flow

where R_{in} and R_{out} are the inner and outer radii of the capillary, respectively.

Figure 6.15 shows the calculated influence of an external radial field on μ_{eo}, for different values of μ_{eo^0}, the electroosmotic mobility in the absence of a radial field. The figure clearly illustrates two empirical findings in EOF-control:

- with increasing external fields, its influence on μ_{eo} decreases;

- when the osmotic mobility without external field is already high, the influence of an external field is smaller than with low μ_{eo^0}.

It may be argued that the possibility to influence the EOF does not imply the possibility to stabilize it; undesired variations resulting from, e.g., adsorption of sample constituents, cannot be eliminated as long as a device for real-time monitoring of the EOF is not available. Still, the phenomenon is interesting in itself, if only since radial fields may occur inadvertently in several experimental situations, possibly affecting the reliability of CE-separations. When a grounded part of the instrumentation is in close proximity of the capillary, it might cause a radial field. Bello et al. [34] explained differences in the EOF, observed with the same capillary installed in different instruments, in this way.

References

[1] A.J. Bard and L.R. Faulkner, "Electrochemical Methods, Fundamentals and Applications", J. Wiley & Sons, New York, 1980, Chapter 12.
[2] I. Fried, "The Chemistry of Electrode Processes"; Academic Press; London, 1973; Chapter 4.
[3] R.J. Hunter, "Foundations of Colloid Science", Clarendon Press; Oxford, 1987; Chapter 6.
[4] J.A. van Lier, P.L. de Bruyn and J.Th.G. Overbeek, J. Phys. Chem. 64 (1960) 1675.
[5] L. Lyklema, J. Electroanal. Chem., 18 (1968) 341.
[6] D.E. Yates, S. Levine and T.W. Healy, Trans. Faraday Soc. 70 (1974) 1807.
[7] J.A. Davis, R.O. James and J.O. Leckie, J. Colloid Interface Sci. 69 (1978) 480.]
[8] H. Poppe, A. Cifuentes and W.Th. Kok, Anal. Chem., 68 (1996) 888.
[9] S. Kaupp, R. Steffen and H. Waetzig, J. Chromatogr. A, 744 (1996) 93.
[10] T.L. Huang, Chromatographia, 35 (1993) 395.
[11] J. Kohr and H. Engelhardt, J. Microcolumn Sep., 3 (1991) 491.
[12] M.F.M. Tavares and V.L. McGuffin, Anal. Chem., 67 (1995) 3687.
[13] J.E. Dickens, J. Gorse, J.A. Everhart and M. Ryan, J. Chromatogr. B, 657 (1994) 401.
[14] M. Mammen, J. Carbeck, E.E. Simanek and G.M. Whitesides, J. Am. Chem. Soc., 119 (1997) 3469.
[15] C. Schwer and E. Kenndler, Anal. Chem., 63 (1991) 1801.
[16] W.R. Jones and P. Jandik, J. Chromatogr., 608 (1992) 385.
[17] J.C. Reijenga, G.U.A. Aben, T.P.E.M. Verheggen and F.M. Everaerts, J. Chromatogr., 260 (1983) 241.
[18] W.R. Jones and J.P. Jandik, J. Chromatogr., 546 (1991) 445.
[19] C.A. Lucy and R.S. Underhill, Anal. Chem., 68 (1996) 300.
[20] A. Emmer, M. Jansson and J. Roeraade, J. Chromatogr., 547 (1991) 544.
[21] K.K.-C. Yeung and C.A. Lucy, Anal. Chem., 69 (1997) 3435.
[22] M. Chen and R.M. Cassidy, J. Chromatogr., 602 (1992) 227.
[23] A. Dougherty, C. Wooley, D. Williams, D. Swaile, R. Cole and M. Sepaniak, J. Liq. Chromatogr., 14 (1991) 907.
[24] G.J.M. Bruin, R. Huisden, J.C. Kraak and H. Poppe, J. Chromatogr., 480 (1989) 339.
[25] S. Hjerten and K. Kubo, Electrophoresis, 14 (1993) 390.
[26] H. Engelhardt and M.A. Cunat-Walter, J. Chromatogr., 716 (1995) 27.
[27] M. Gilges, M.H. Kleemiss, G. Schomburg, Anal. Chem., 66 (1994) 2038.
[28] H. Bayer and H. Engelhardt, J. Microcolumn Sep., 8 (1996) 479.
[29] M.W.F. Nielen, J. High Resolut. Chromatogr., 16 (1993) 62.
[30] C.S. Lee, W.C. Blanchard, C.-T. Wu, Anal. Chem. 63 (1990) 1550.
[31] M.A. Hayes and A.G. Ewing, Anal. Chem. 64 (1992) 512.
[32] C.A. Keely, R.R. Holloway;, T.A.A.M. van de Goor and D. McManigill, J. Chromatogr.A, 652 (1993) 283.
[33] C.S. Lee, C.-T. Wu, T. Lopes and B. Patel, J. Chromatogr., 559 (1991) 133.
[34] M.S. Bello, L. Capelli and P.G. Righetti, J. Chromatogr.A 684 (1994) 311

7 The Background Electrolyte

7.1 The role of the background electrolyte in CZE

In capillary zone electrophoresis (CZE), which is the most general and widely applied mode of CE, analyte ions are separated as zones migrating with different velocities through a so-called background electrolyte (BGE), a solution of a buffering salt. It is sometimes (inaccurately) stated that the separation medium should contain a salt to make the electrophoretic current possible; however, this current is not essential for the differential migration of analyte ions under the influence of the electric field applied. The role of the BGE is to provide constant conditions for the analyte ions during the separation, and to make these conditions independent of the sample composition. Only under such conditions one can expect to be able to identify a sample component by its migration time in CZE.

Since the velocity of an analyte zone is the product of the effective mobility of the component (μ_{eff}) and the local electric field strength (E), the BGE is expected to keep these two factors constant. Keeping the effective mobility of an ionogenic compound constant implies to keep the solution pH constant, independent of the analyte concentration in the zone. Therefore, a buffering solution should be used as the BGE. This is not only good practice for the separation of weakly acidic or basic compounds (for which the effective mobility will be a function of the pH), but also for the separation of strong ions. In the latter case the pH should be kept constant and stable to have a constant electroosmotic mobility from run to run.

The presence of analyte ions in a zone somewhere along the capillary may influence the local electric field strength at that position. This is clear when one regards the electrophoretic current as a function of the local field strength and the conductivity κ of the solution:

$$i = \frac{\pi}{4}\, d_c^2 \cdot \kappa\,(x) \cdot E(x) \qquad (7.1)$$

Over the length of the capillary the current will be constant. Therefore, the local field strength will be decreased at a position (in a zone) with a conductivity higher than that of the blank BGE and *vice versa*. The presence of analyte ions in a zone will almost inevitably change the local conductivity to some extent; the difference of the conductivity from that of the blank BGE will increase with the concentration of the analyte. This makes the velocity of an analyte zone in principle a function of its concentration. By using a suitable salt solution as the BGE one tries to keep this effect as small as possible.

The dependency of the local field strength on the concentration of the analyte ion within a zone may lead to the broadening of the zone by a process referred to as electromigration dispersion. This phenomenon is illustrated in Figure 7.1. First assume that the conductivity

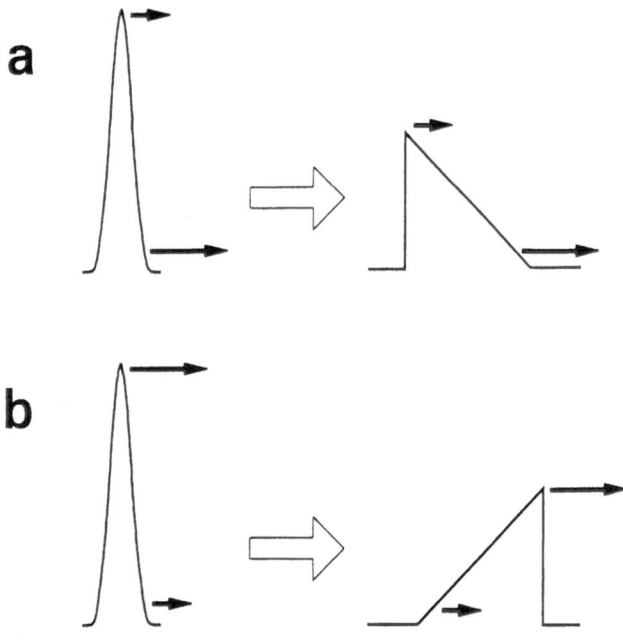

Figure 7.1
Zone broadening and peak deformation when the migration velocity decreases (a) or increases (b) with the concentration of the analyte.

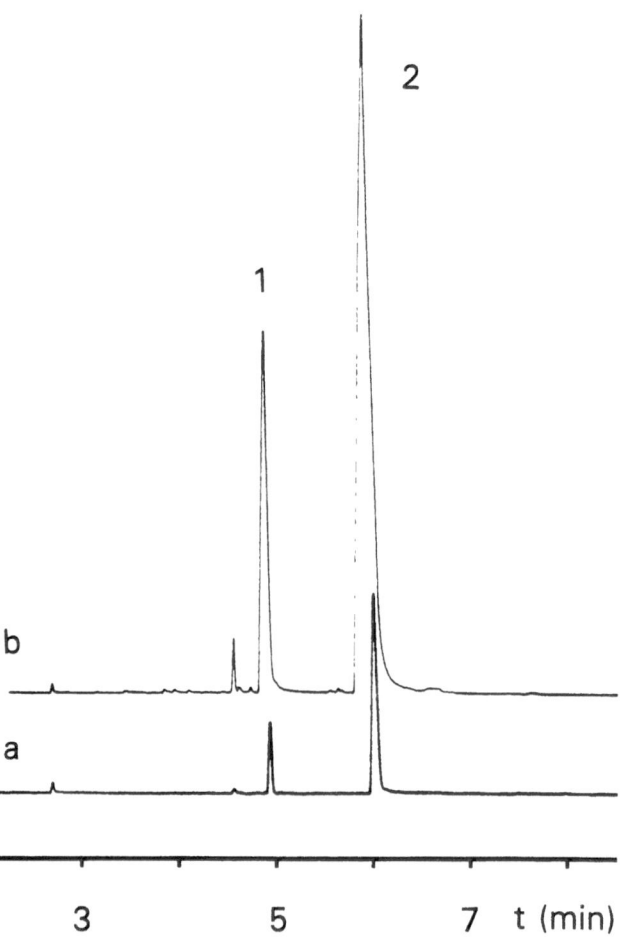

Figure 7.2
Separation of 4-naphtalene sulphonic acid (1) and 3,6-naphtalene disulphonic acid (2) in 10 mmol L^{-1} borate buffer. Sample concentration: (a) 10^{-4} and (b) 10^{-3} mol L^{-1}.

0009-5893/00   S-36-08 $ 03.00/0

of the solution increases with increasing analyte ion concentration (Figure 7.1 a). In this situation the field strength, and with that the electrophoretic velocity of the analyte ions, will be relatively low in a part of the zone with a high analyte concentration. A zone with an originally symmetric (Gaussian) concentration profile will be deformed, since the front end will migrate with a higher velocity than the peak top, while the tail of the original symmetric peak will catch up with the top. In this way a triangular concentration profile will develop, with a sloping front and a steep backside. During electrophoresis, the distance between the fast moving front end and the slower moving peak maximum will steadily increase. The situation in which the solution conductivity decreases with the analyte concentration is sketched in Figure 7.1b. Now the front end of the zone is overhauled by the peak maximum, while the tail lags further and further behind during electrophoresis. A triangular peak with a sharp front and a sloping backside will develop. In both cases the result is that zones will get broadened.

Electromigration dispersion is a general phenomenon in CZE. Its effect is more pronounced with increasing sample amounts; therefore it is also called overloading. An experimental example is given in Figure 7.2, which shows the separation of a number of naphthalene sulphonic acids with a borate buffer as BGE. With a low sample concentration (10^{-4} mol L^{-1}), peaks are narrow and more or less symmetrical. With a high sample concentration (10^{-3} mol L^{-1}) much wider, triangular peaks are observed. Overloading or electromigration dispersion can generally be recognised by the following characteristics:

- with increasing analyte amount symmetric peaks are deformed into tailing or fronting triangular peaks; in one electropherogram both tailing and fronting peaks may be observed;
- the width of a triangular peak increases with the amount of analyte injected;
- peak heights do not increase proportional with the amount of analyte injected; this is of course related to the increasing peak widths;
- the migration times of the peak tops are slightly shifted with increasing injected amount;
- the overloading effects decrease when a higher salt concentration in the BGE is used.

In Figure 7.3 the effects of overloading for the separation of Figure 7.2 are shown. While the migration time of the peak maximum shifts with increasing analyte concentration, that of the peak front (where the extrapolation of the sloping side crosses the baseline) is fairly constant. By definition the analyte concentration at the front (or tail) end of a peak approaches zero; the velocity of this side of the peak is therefore independent of the total amount of analyte in the zone.

The triangular shape of overloaded peaks suggests that the velocity of analyte ions increase or decrease in a

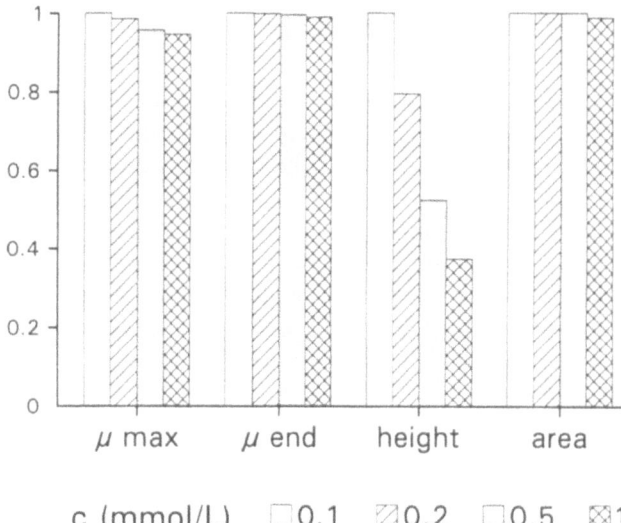

Figure 7.3

Dependency of separation parameters on the sample concentration. Conditions as in Figure 7.2; sample concentrations in mmol L^{-1}. Peak height and area relative to sample concentration.

linear fashion with their concentration. Together with the observation that the degree of overloading of a peak decreases inversely proportional with the salt concentration in the BGE, this leads to the following equation to describe overloading in CZE:

$$v_i(c) = v_{i,o} \cdot \left(1 + \beta_{EMD} \cdot \frac{c_i}{c_{BGE}} \right) \qquad (7.2)$$

where $v_i(c)$ is the velocity of a part of a zone with a particular analyte concentration c_i, $v_{i,0}$ the velocity of the front or tail of the zone (with $c_i = 0$), and c_{BGE} the salt concentration of the BGE. The factor β_{EMD} can be regarded as a constant for a particular analyte-ion/BGE-type combination. With a positive value for β_{EMD} the velocity of the analyte ions increases with their concentration; this will lead to a peak with a sharp front and a tailing backside. With a negative value for β_{EMD} a fronting peak with a steep backside is expected. The steepness of the sloping side of a peak is expected to be inversely proportional to β_{EMD}; that is, with a low value of β_{EMD} the differences in velocity for the two sides of the peak are expected to be small.

From Equation 7.2 it will be clear that one can do the following to diminish zone broadening by overloading:

- dilute the sample; this is of course limited by the requirements of detectability, especially when components with high and low concentrations have to be determined in the same sample;
- increase the BGE salt concentration; here, limits will be set by the Joule heating in the capillary;
- find a BGE type giving a low electromigration constant (β_{EMD}) for critical analyte zones.

The first two solutions for the overloading problem are trivial. One has to look at principle of electromigration dispersion in some more detail to predict what BGE type will be give a low β_{EMD} value for a certain analyte ion.

7.2 The Kohlrausch regulating function

Many important features of CE, such as electromigration dispersion, sample stacking and indirect detection, can be predicted theoretically with the help of moving-boundary equations, that describe mass transport during electrophoresis. Already in the previous century Kohlrausch derived an elegant solution of the moving boundary equations for systems containing multiple ionic species [1]. Kohlrausch named the equation derived the "beharrliche Function" or regulating function. Since this function plays an essential role in CE, Here, the Kohlrausch regulating function will be derived for a very simple system of ions in solution. The result is also applicable for more complicated solutions, but for this generalisation the reader should consult the original literature [2, 3].

Assume that only three ions are present in the solution in the capillary: the analyte ion A^+, the background electrolyte ion B^+, and a counter ion C^-. Their concentrations will be denoted as c_A, c_B and c_C, respectively. The concentration of C^- follows directly from the concentrations of A^+ and B^+, since the electroneutrality principle requires that $c_C = c_A + c_B$. Assume now that at the start of an experiment a boundary plane over which the composition of the solution differs, moves through the capillary. We can discern two homogeneous phases α and β, separated by the boundary (Figure 7.4). The concentrations of the ions in phase α are denoted as c_A^α, c_B^α and c_C^α, and those in phase β as c_A^β, c_B^β and c_C^β. The boundary may be the front of the analyte zone (when $c_A^\beta = 0$), the backside of it (when $c_A^\alpha = 0$), or it may indicate any change in concentration of A^+ and/or B^+.

From $t=0$, an electric field is applied for one unit of time, with the negative electrode at the right. A^+ and B^+ ions from phase α will migrate to the right, thereby moving the boundary to the right. At $t=1$ the boundary may be located at a distance S from the original one. The mass fluxes of A^+, B^+ and C^- will be regarded over two planes P and Q that are located just outside the original and new boundary positions. The position of these planes is chosen in this way because P is in phase α and Q in phase β during the whole experiment. To simplify calculations the cross-section of the capillary is supposed to be 1 unit of area.

The mass flux of A^+ over plane P, entering part S of the capillary, in one unit of time, is dependent of the field strength in phase α:

$$\Phi_{A,in} = \mu_A \cdot E^\alpha \cdot c_A^\alpha \tag{7.3}$$

The mass flux of A^+ over plane Q, leaving part S, depends on the field strength in phase β (E^β) which is not necessarily the same as E^α:

$$\phi_{A,out} = \mu_A \cdot E^\beta \cdot c_A^\beta \tag{7.4}$$

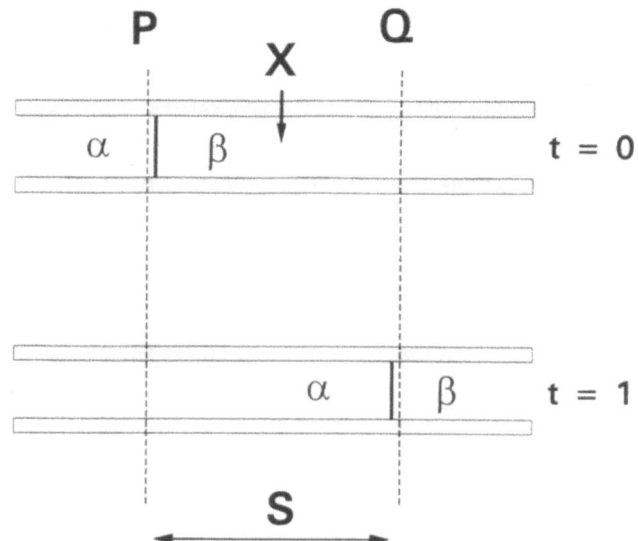

Figure 7.4
Scheme of the moving boundary in the Kohlrausch problem.

If we divide the difference of the mass fluxes in and out by the volume of part S, we find the change in the concentration of A^+, when phase β is replaced by phase α:

$$c_A^\alpha - c_A^\beta = \frac{1}{S} \cdot \mu_A \cdot (E^\alpha \cdot c_A^\alpha - E^\beta \cdot c_A^\beta) \tag{7.5}$$

Division of both sides of this equation by μ_A gives:

$$\frac{c_A^\alpha}{\mu_A} - \frac{c_A^\beta}{\mu_A} = \frac{1}{S} \cdot (E^\alpha \cdot c_A^\alpha - E^\beta \cdot c_A^\beta) \tag{7.6}$$

A similar computation can be made for B^+:

$$\frac{c_B^\alpha}{\mu_B} - \frac{c_B^\beta}{\mu_B} = \frac{1}{S} \cdot (E^\alpha \cdot c_B^\alpha - E^\beta \cdot c_B^\beta) \tag{7.7}$$

Since ion C^- migrates in the other direction, the terms for the mass fluxes in and out should be interchanged; the equivalent equation for C^- is therefore:

$$\frac{c_C^\alpha}{\mu_C} - \frac{c_C^\beta}{\mu_C} = -\frac{1}{S} \cdot (E^\alpha \cdot c_C^\alpha - E^\beta \cdot c_C^\beta) \tag{7.8}$$

Let us now add the three equations 7.6–7.8:

$$\frac{c_A^\alpha}{\mu_A} + \frac{c_B^\alpha}{\mu_B} + \frac{c_C^\alpha}{\mu_C} - \frac{c_A^\beta}{\mu_A} - \frac{c_B^\beta}{\mu_B} - \frac{c_C^\beta}{\mu_C}$$

$$= \frac{1}{S} \cdot \{E^\alpha \cdot (c_A^\alpha + c_b^\alpha - c_C^\alpha) -$$

$$E^\beta \cdot (c_A^\beta + c_B^\beta - c_C^\beta)\} \tag{7.9}$$

Because of the electroneutrality principle, $(c_A + c_B - c_C)$ is zero for both phases. The right-hand side of Equation 7.9 is therefore zero, irrespective of the field strength in the two phases. We finally find the Kohlrausch equation:

$$\frac{c_A^\alpha}{\mu_A} + \frac{c_B^\alpha}{\mu_B} + \frac{c_C^\alpha}{\mu_C} = \frac{c_A^\beta}{\mu_A} + \frac{c_B^\beta}{\mu_B} + \frac{c_C^\beta}{\mu_C} \qquad (7.10)$$

or more generally:

$$\sum \frac{c_i}{\mu_i} = \omega = \text{constant} \qquad (7.11)$$

The constant sum ω is called the Kohlrausch value of the solution.

Let us now recapitulate what has happened at observation point X. During electrophoresis a zone boundary has passed this point. The concentrations of A, B and/or C have changed, but the regulating function ω has not changed. When no boundary passes, ω will certainly not change. Therefore we may conclude that during electrophoresis ω will remain constant at any point in the solution.

The general practice in CZE is that the whole capillary, except for the capillary tip where the sample is introduced, is filled with one BGE, with one specific value of ω. During the separation zones will migrate through the solution, where originally (blank) BGE was present. Therefore, all zones will have the Kohlrausch value of the BGE. The sample plug may possibly have another ω value. During the separation this deviating value of the regulating function will remain at the same position in the solution. Since the whole solution moves by electro-osmosis, the deviation is swept along the capillary with the osmotic velocity.

The Kohlrausch function was derived above for singly charged, strong ions. However, Equation 7.11 is also valid:

- when multiple charged, strong ions are present in the solution; the concentrations in Equation 7.11 should then be replaced by the equivalent concentrations ($|z_i|c_i$);
- when weakly acidic or basic compounds, with a single ionic charge, are present in the solution; in this case the concentrations in Equation 7.11 should be interpreted as the total, analytical concentrations of the compounds in question [4].

7.3 Overloading with strong analyte ions

In first approximation it can be assumed that an analyte zone migrates with a velocity determined by the ionic mobility and the field strength in the background electrolyte, to be calculated from the applied voltage V_{appl} and the capillary length L. However, as will be shown, the presence of the analyte ions influences the local field strength in the zone. In this way the velocity of the zone depends on the analyte ion concentration. The Kohlrausch regulating function is essential in understanding this phenomenon.

Assume an analyte zone ζ with analyte concentration c_A^ζ, some time after the application of the electric field. The

zone ζ has left the original sample zone completely, and is now at a position previously occupied by the pure background electrolyte β. Kohlrausch's rule states that the regulating function ω of zone ζ must be equal to that of the background electrolyte in this place before the separation:

$$\frac{c_A^\zeta}{\mu_A} + \frac{c_B^\zeta}{\mu_B} + \frac{c_C^\zeta}{\mu_C} = \frac{c_B^\beta}{\mu_B} + \frac{c_C^\beta}{\mu_C} \qquad (7.12)$$

$$\left(\frac{1}{\mu_A} + \frac{1}{\mu_C}\right) \cdot c_A^\zeta + \left(\frac{1}{\mu_B} + \frac{1}{\mu_C}\right) \cdot c_B^\zeta$$

$$= \left(\frac{1}{\mu_B} + \frac{1}{\mu_C}\right) \cdot c_B^\beta \qquad (7.13)$$

Rearrangement gives an expression for c_B^ζ that we will need later:

$$c_B^\zeta = c_B^\beta - \frac{\mu_B}{\mu_A} \cdot \frac{\mu_A + \mu_C}{\mu_B + \mu_C} \cdot c_A^\zeta \qquad (7.14)$$

Now let us compare the field strength in the analyte zone ζ with that in the separation buffer β. Since the largest part of the capillary will still be containing the (pure) background electrolyte, E^β is still approximately given by V_{appl}/L. From the constancy of the current density in the capillary we have:

$$E^\zeta = \frac{\kappa^\beta}{\kappa^\zeta} \cdot E^\beta \qquad (7.15)$$

Substituting the expressions for the conductivities:

$$E^\zeta = \frac{(\mu_B + \mu_C) \cdot c_B^\beta}{(\mu_B + \mu_C) \cdot c_B^\zeta + (\mu_A + \mu_C) \cdot c_A^\zeta} \cdot E^\beta \qquad (7.16)$$

The value of c_B^ζ was derived before; substitution and some rearrangements give:

$$E^\zeta = \frac{1}{1 - \left(\dfrac{\mu_B}{\mu_A} - 1\right) \cdot \dfrac{\mu_A + \mu_C}{\mu_B + \mu_C} \cdot \dfrac{c_A^\zeta}{c_B^\beta}} \cdot E^\beta \qquad (7.17)$$

This equation shows that the actual field strength in the zone, and hence its velocity, depends on the concentration of the analyte, or, more precisely, on the ratio of this concentration and that of the background electrolyte.

To evaluate the meaning of the equation for the peak shapes in CE, we can simplify it somewhat when it is assumed that the analyte concentration is much smaller than the buffer concentration. Then we can write:

$$E^\zeta = \frac{1}{1 - \beta_{EMD} \cdot \dfrac{c_A^\zeta}{c_B^\beta}} \cdot E^\beta \approx \left(1 + \beta_{EMD} \cdot \frac{c_A^\zeta}{c_B^\beta}\right) \cdot E^\beta \qquad (7.18)$$

where the EMD-constant β_{EMD} is given by:

$$\beta_{EMD} = \left(\frac{\mu_B}{\mu_A} - 1 \right) \cdot \frac{\mu_A + \mu_C}{\mu_B + \mu_C} \qquad (7.19)$$

Only when the analyte is present in very low concentrations ($c_A^{\vee} \to 0$), the field strength in the analyte zone is the same as in the buffer. The zone than migrates with the 'expected' velocity $\mu_A \cdot E^\beta$. For higher analyte concentrations a deviating velocity will be observed. Since concentration gradients are developed within the analyte zone, e.g. by diffusion, this non-ideal migration behaviour causes zone-broadening. A zone can be thought of as being built up from different narrow zones with different concentrations. To simplify the discussion, a peak can be divided in a front and tail with low concentration, and a top with a high concentration of analyte ions. Three different situations can be observed.

(a) $\mu_A = \mu_B$; $\beta_{EMD} = 0$

When the mobilities of the analyte ion and the background ion (with the same sign) exactly match each other, the field strength in the analyte zone is equal to the 'mean' field strength in the buffer; the velocity of the zone does not depend on the analyte concentration. The top and flanks of an analyte peak travel with the same velocity, and only the usual zone broadening is observed [5, 6].

(b) $\mu_A > \mu_B$; $\beta_{EMD} < 0$

When the mobility of the analyte ion is higher than that of the background ion, the zone velocity decreases when the analyte concentration is increased. This means that the front and the tail of a peak migrate faster than the top. A strongly fronting peak will be observed with an abrupt backside.

(c) $\mu_A < \mu_B$; $\beta_{EMD} > 0$

In this case the top of a peak migrates faster than the front and the tail. Tailing peaks with a steep frontside will be observed.

Figure 7.5 shows schematically the peak shapes that will be observed in the separation of a mixture of positive and negative ions. In the interpretation of this figure one should keep in mind that the zone velocities are superimposed on the electroosmotic velocity. Fronting and tailing of a peak is defined relative to the solution; the direction of migration can be seen as away from the electroosmotic (neutral) peak.

When non-ideal migration is the main cause of zone-broadening, the front or tail of the observed peaks should be taken as indicative for the mobility of the ions. As was already shown in Figure 7.2, the migration time of the peak end is much less dependent on the sample concentration than that of the peak maximum.

Under overloading conditions, the peak widths increase approximately proportional with the square root of the

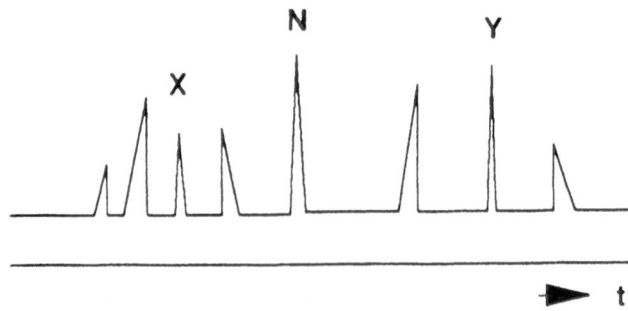

Figure 7.5
Virtual electropherogram showing overloading for strongly anionic and cationic analytes. N: neutral peak indicating the EOF; X: symmetric peak for anion with the same mobility as the BGE anion; Y: symmetric peak for cation with the same mobility as the BGE cation.

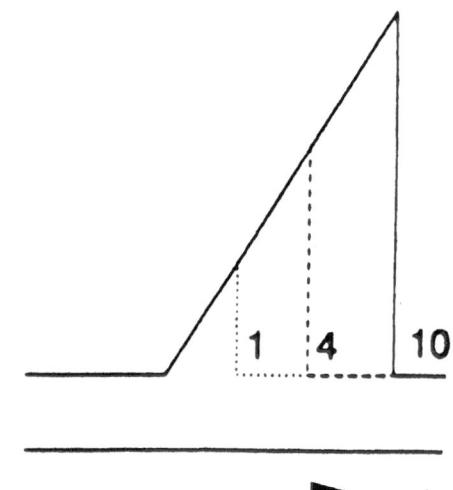

Figure 7.6
Influence of the injected sample amount on the position of the peak maximum and the height and area of an overloaded peak. The numbers in the figure indicate the relative amounts injected.

amount injected (see Figure 7.6). This implies that peak heights are also increasing with the square root of the amount injected only.

7.4 The quality of the BGE

In the previous paragraphs it has been shown how the dependency of the local field strength on the concentration of an analyte ion in a zone can lead to the broadening of the zone. With a certain BGE concentration, zone widths increase in proportion to the square root of the amount of analyte injected. The ratio of the analyte and the BGE concentrations determines the extent of this zone broadening. The susceptibility of an analyte ion/BGE combination to overloading can be expressed in the electromigration dispersion constant β_{EMD}. With the help of the Kohlrausch function a simple

The Background Electrolyte

formula relating β_{EMD} to ionic mobilities has been derived for strong ions.

Overloading can be avoided for one particular (strong) analyte ion by choosing a BGE co-ion (having the same sign as the analyte ion) that matches the mobility of the analyte ion exactly. Such an approach is useful when there is one particular pair of compounds in the sample that is difficult to separate; with this pair overloading is the most critical for the resolution. This "matching rule" can also be applied when one of the compounds in the sample is present in much higher concentration than the others; an example could be in purity studies of (pharmaceutical) preparations. However, in many cases analytes in a sample have strongly differing mobilities, so that for most of the zones overloading to some extent will be unavoidable. Figure 7.7A shows the electromigration dispersion constant as a function of the analyte mobility for a number of co-ion mobilities. For every co-ion there is one matching analyte mobility for which overloading will be absent. Analytes with a lower mobility than the co-ion have a positive β_{EMD} (tailing peaks), those with a higher mobility a negative β_{EMD} (fronting peaks). The curves in the figure show that tailing is more readily severe than fronting when the mobilities do not match exactly.

The influence of the counter-ion mobility (the BGE ion with a sign opposite to that of the analyte) on overloading is small. In Figure 7.7B the value of β_{EMD} is shown as a function of the analyte mobility, with a co-ion mobility of 25 and counter-ion mobilities of 10 and 50 ($\times 10^{-9}\,\mathrm{m^2\,V^{-1}\,s^{-1}}$). The use of a "fast" counter-ion is to be preferred for fast analytes, and a slow counter-ion for analytes with a mobility lower than that of the co-ion, but the differences are small.

From Figure 7.7 it can be seen that when a wide range of analyte mobilities have to be handled, values of β_{EMD} are easily as large as ± 0.5. This sets severe limits to the analyte concentrations that can be introduced in the CE system without serious overloading. To preserve the high plate numbers typical for CE, in general the analyte concentration at the peak top should be kept below 1 % of the BGE salt concentration. With a typical BGE concentration of 10 mmol L^{-1}, the maximum analyte concentration is in the order of 10^{-4} mol L^{-1}.

Increasing the BGE concentration is an obvious way to oppose overloading. However, as has been mentioned before, the BGE concentration can not be raised above a certain limit for reasons of excessive Joule heating during electrophoresis. The Joule heating is proportional to the conductivity of the BGE. Therefore, a BGE salt type giving a low conductance might be preferable in this respect. To compare different BGE types in terms of susceptibility for overloading by a certain analyte ion, both the electromigration dispersion constant and the equivalent conductance should be considered. An electromigration dispersion quality factor can be defined for a BGE as:

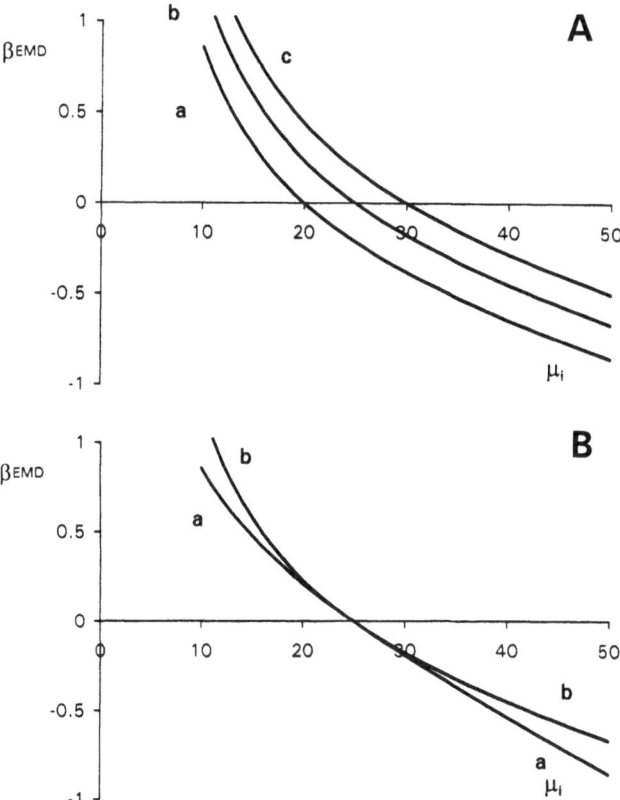

Figure 7.7

The EMD-constant as a function of the analyte mobility. (A): Influence of the mobility of the co-ion in the BGE; $\mu_B = 20$ (a), 25 (b) or 30 (c); $\mu_C = 50$. (B): Influence of the mobility of the counter-ion in the BGE; $\mu_C = 10$ (a) or 50 (b); $\mu_B = 25$.

$$Q_{EMD} = \left| \frac{F}{\beta_{EMD} \cdot \lambda} \right| \qquad (7.20)$$

where λ is the equivalent conductance of the BGE salt. The Faraday constant has been introduced in this definition for simplicity; for a simple salt B^+C^- λ can be expressed as $F(\mu_B + \mu_C)$. By combining Equation 7.25 with the expression for β_{EMD} (Equation 7.24) one finds a simple formula to express the quality of a BGE type for strong analyte ions:

$$Q_{EMD}\ \frac{\mu_A}{|\mu_B - \mu_A| \cdot (\mu_A + \mu_C)} \qquad (7.21)$$

when A is the analyte ion, B the co-ion and C the counter-ion in the BGE. Figure 7.8 shows the dependency of the BGE quality factor on the mobility of the analyte; in Figure 7.8A this relation is shown for different co-ion mobilities; in Figure 7.8B the influence of the counter-ion is shown. The use of low conductivity buffer salts, such as those of the well known biological buffers, has often been advocated for CE. From Figure 7.8 it appears that such buffering salts do not specifically exhibit a better quality in respect to electromigration dispersion. For analyte ions with a high mobility a fast BGE co-ion should be chosen. Only for the counter-ion in the BGE it

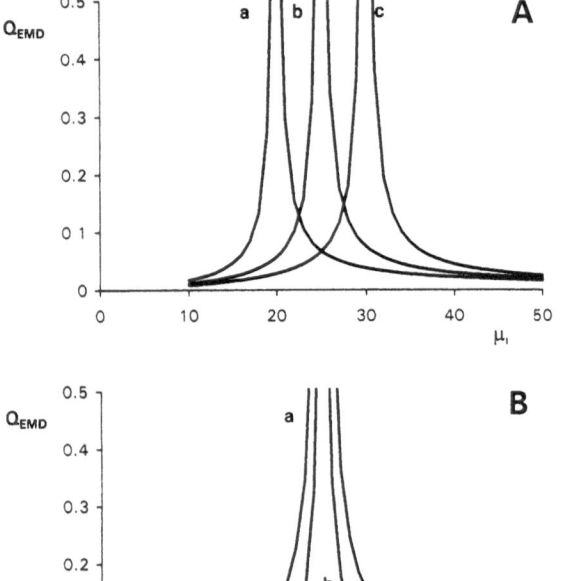

Figure 7.8

The quality parameter of the BGE as a function of the analyte mobility. (A): Influence of the mobility of the co-ion in the BGE; μ_B = 20 (a), 25 (b) or 30 (c); μ_C = 50. (B): Influence of the mobility of the counter-ion in the BGE; μ_C = 10 (a) or 50 (b); μ_B = 25.

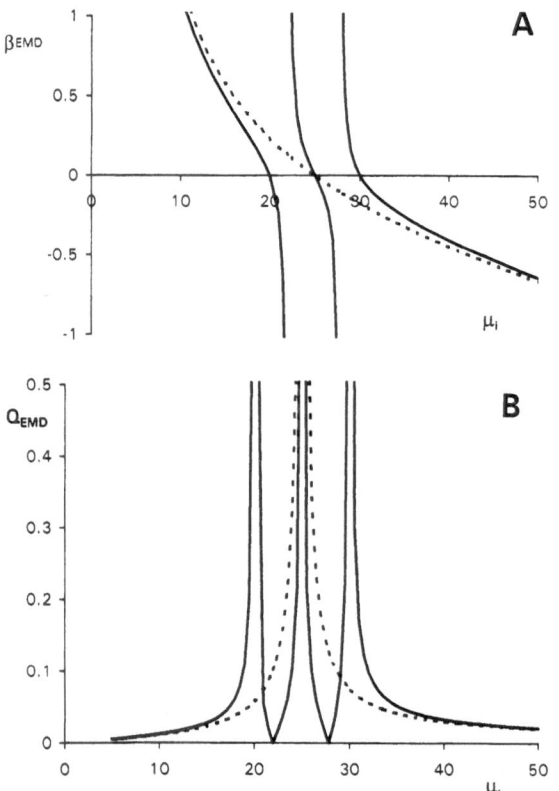

Figure 7.9

The EMD-constant and the quality parameter with a mixed BGE as a function of the analyte mobility. The BGE contains three co-ions in equal concentrations, with mobilities of 20, 25 and 30; the counter-ion mobility is 50. The dotted lines give the β_{EMD}- and Q_{EMD}-values for a BGE with a single co-ion (with a mobility of 25).

can be stated in general that this should have a low mobility. With a slow counter ion the range of analyte mobilities around that of the BGE co-ion, for which low electromigration dispersion will be observed, is wider.

It has been proposed to use mixed BGE's to oppose overloading for a wider range of analyte mobilities [7]. In Figure 7.9 the effect of this approach is shown. The EMD constant and the quality parameter for a mixed BGE, containing three different co-ions, can be compared with those of a single salt BGE, with an equal total salt concentration. The figure shows the following:

- with a mixed BGE, a zero EMD constant and thus a high BGE quality is obtained when an analyte ion matches any of the BGE co-ions in mobility;

- with a mixed BGE however, this matching of mobilities should be more precise than with a single salt BGE; the "good quality" regions for the analyte mobility are narrower;

- with a mixed BGE for some analyte mobilities an extremely high susceptibility for overloading will be found; this is at positions in the electropherogram close to so-called system peaks (see paragraph 11.2).

Because of the existence of mobility regions with zero quality, the use of mixed BGE's to oppose electromigration dispersion can not be recommended in general.

7.5 Overloading with weak ions

As has been shown in the previous paragraph it is relatively easy to predict the degree of overloading for strong ions. When the composition of the BGE is simple, the electromigration dispersion constant for a particular analyte ion can be obtained directly from the mobilities of the ions in question. For weakly acidic or basic analytes, the prediction of their overloading characteristics is much more complicated [8, 9, 10]. For such compounds overloading is related to two different non-idealities:

- as with strong ions, the presence of analyte ions in a zone may cause a deviating conductivity, and thereby a deviating local electric field strength;

- weakly acidic or basic analytes will compete with the buffering BGE component for the protons in the solution; the presence of such an analyte may cause a shift of the proton equilibrium of the BGE, and the pH of a zone may be different from that in the blank BGE [11, 12].

Both effects will increase with increasing analyte concentration, and therefore contribute to the deformation and broadening of zones. In some cases the two effects reinforce each other, while in other cases they may cancel each other out. This may result in a seemingly haphazard sequence of fronting and tailing peaks in an electropherogram. An example is given in Figure 7.9, which shows the separation of a number of weak and strong organic acids in a 4-aminobenzoic acid buffer.

The conductivity effect for weak ions is similar to that for strong ions. For a certain analyte concentration in a

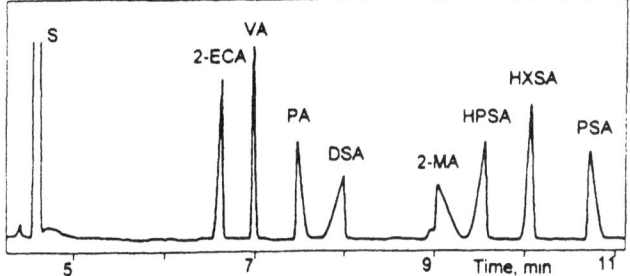

Figure 7.10
The separation of a number of weak and strong acids with a 10 mmol L⁻¹ 4-aminobenzoic acid buffer as BGE. Sample concentration: 0.5 mmol L⁻¹ for each compound. Indirect UV detection at 267 nm. Reproduced from ref. [10].

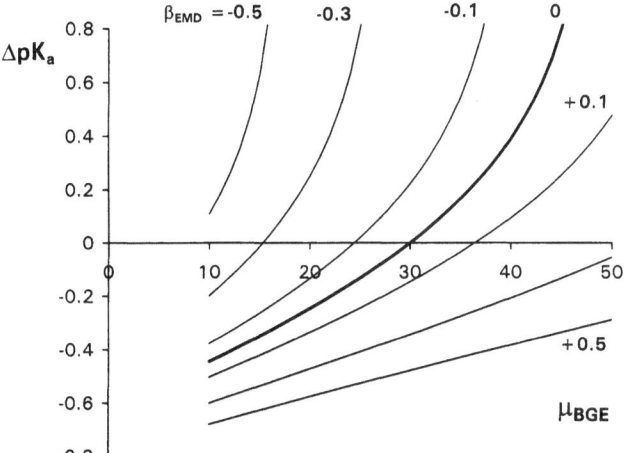

Figure 7.11
The EMD-constant for a weakly acidic analyte with a mobility of 30 and an ionisation degree of 50%, as a function of the pK_A and ionic mobility (μ_{BGE}) of the BGE compound.

obtained describing the conductivity and pH-shift contributions to the electromigration dispersion constant, but these expressions can be fairly complicated [10]. For instance, for a HA/A⁻ type analyte in a HB/B⁻ type BGE buffer, β_{EMD} can be written as:

$$\beta_{EMD} = \alpha_A \cdot \frac{\mu_B - \mu_A}{\mu_A} \cdot \frac{\mu_A + \mu_C}{\mu_B + \mu_C} + (\alpha_B - \alpha_A) \cdot \frac{1 - \alpha_A}{1 - \alpha_B} \quad (7.22)$$

where α_A and α_B are the degrees of ionisation (at zero analyte concentration) of the analyte and BGE component, respectively. The values of α are of course determined by the pK_A-values for both compounds and the acid-base composition of the BGE.

In some but not all cases one can try to find a BGE buffer type with such a pK_A and ionic mobility that the two overloading effects cancel each other out for a specific analyte. Of course the ionisation degree of the analyte, and the pH of the BGE, should be determined in first instance by the requirement to separate the analyte from interfering substances. However, for a given separation pH there may be a choice of buffer types of which one has the highest quality in respect to overloading. Figure 7.11 shows values of the EMD-constant for an analyte with an ionic mobility of 25×10^{-9} m² V⁻¹ s⁻¹ and an ionisation degree of 50 %, as a function of the pK_A and mobility of the BGE compound. Any buffer type with properties represented by the thicker line (with $\beta_{EMD} = 0$) in the figure will show the best performance against overloading.

References

[1] F. Kohlrausch, Ann. Phys. Chem., 62 (1897) 14.
[2] V.P. Dole, J. Am. Chem. Soc., 67 (1945) 1119.
[3] L.G. Longworth, J. Phys. Chem., 51 (1947) 171
[4] L.M. Hjelmeland and A. Crambach, Electrophoresis, 3 (1982) 9.
[5] S. Hjerten, in G. Milazzo (Ed.), Topics in Bioelectrochemistry and Bioenergetics, Vol. 2, Wiley, New York, p. 106.
[6] F.E.P. Mikkers, F.M. Everaerts, and Th.P.E.M. Verheggen, J. Chromatogr., 169 (1979) 1.
[7] T. Wang and R.A. Hartwick, J. Chromatogr., 589 (1992) 307.
[8] V. Sustacek, F. Foret, and P. Bocek, J. Chromatogr., 545 (1991) 239.
[9] J.L. Beckers, J. Chromatogr. A, 693 (1995) 347.
[10] X. Xu, W.Th. Kok and H. Poppe, J. Chromatogr. A, 742 (1996) 211.
[11] S.V. Ermakov, O.S. Mazhorova, and M.Y. Zhukov, Electrophoresis, 13 (1992) 838.
[12] S.V. Ermakov and P.G. Righetti, J. Chromatogr. A, 667 (1994) 257.

zone, the BGE concentration is established by the Kohlrausch rule. The resulting solution conductivity may be different from that of the blank BGE. The relation between the analyte concentration and the local field strength is in this case more complicated than for strong ions (Equation 7.22), because the degree of ionisation of the analyte and the BGE species plays a role. The magnitude of the pH-shift in an analyte zone depends on the mobilities of the ions involved as well as on the pK_a-values of the analyte and the buffering BGE constituent. Approximate analytical expressions can be

8 Sample Introduction

8.1 Hydrodynamic and electrokinetic injection

Basically there are two different methods to introduce the sample into the first small part of the capillary: hydrodynamic (driven by a pressure difference over the capillary) and electrokinetic. The differences between the two methods will be discussed.

When a pressure difference is applied between the inlet of the capillary in the sample solution and the outlet, a certain volume of the sample solution will flow into the capillary. The pressure difference may be accomplished by gas pressure on the sample vial, by vacuum suction at the capillary end or by applying a height difference between the inlet and outlet. The (average) length L_{inj} of the sample zone introduced is given by:

$$L_{\mathrm{inj}} = \frac{d_c^2}{32\eta \cdot L} \cdot P \cdot t_{\mathrm{inj}} \qquad (8.1)$$

where d_c and L are the diameter and (total) length of the capillary, η the viscosity of the BGE solution, P the pressure difference and t_{inj} the time during which the pressure is applied. For aqueous BGE solutions at 25 °C, the injection zone length can be calculated as:

$$L_{\mathrm{inj}} = 3.5 \times 10^6 \cdot \frac{d_c^2}{L} \cdot P \cdot t_{\mathrm{inj}}$$

The numerical value is valid when the lengths and diameter are expressed in meter, the pressure in bar and the time in seconds. When a height difference between inlet and outlet is applied for sample injection, the pressure is 1 mbar for a height difference of 1 cm.

The length of the sample zone with hydrodynamic injection is dependent on the viscosity of the solution inside the capillary during injection, i.e., on the viscosity of the BGE. This parameter is easily kept constant when the capillary can be thermostatted. The viscosity of the sample solution is of minor importance, since it occupies only a very short part of the capillary length. Therefore, the repeatability of hydrodynamic injection can be excellent even when large viscosity differences between samples exist. In routine applications a precision of <1 % is easily achieved [1]. In an extensive study on the precision of CE with hydrodynamic injection [2] it has been found that for a good precision it is important:

- to keep the temperature constant;
- to avoid very short (1 – 2 s) injection times;
- to be aware of buffer depletion (see paragraph 4.3);
- to use an internal standard.

Since the sensitivity in CE is often problematic, one usually tries to inject as much sample as possible. With hydrodynamic injection, the sample flows into the capillary with a parabolic flow profile (see Figure 8.1 a). Therefore, the sample zone is spread out over a piece of

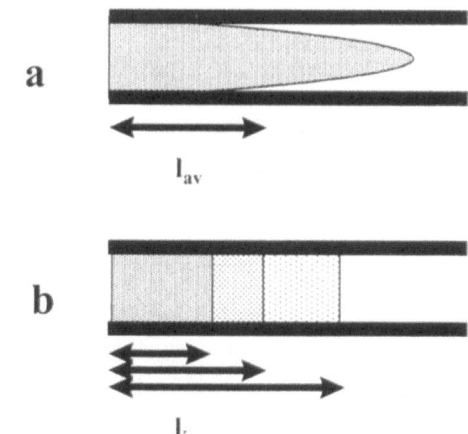

Figure 8.1
Sample zone profiles after (a) hydrodynamic and (b) electrokinetic injection.

capillary of twice its "average" length. This should be kept in mind when the contribution of the sample introduction to zone-broadening is discussed. The standard deviation in length units describing the injected plug (σ_z) is approximately 0.5 times the average length L_{inj}. From the plate number N required for the separation at hand, a maximum value for L_{inj} is easily calculated:

$$L_{\mathrm{inj}} < 2L / \sqrt{N} \qquad (8.2)$$

For a particular capillary the maximum product of injection pressure and time is obtained with the equations above.

Alternatively the sample can be introduced electrokinetically, by applying a voltage for a short time with the inlet of the capillary situated in the sample vial. Usually a lower voltage is applied during injection than during the actual separation, to improve the precision of the injection. The sample enters the capillary by a combined effect of electroosmosis and electrophoresis. This implies that the amount of an analyte ion injected depends on its mobility. An analyte with a mobility in the same direction as the osmotic flow occupies a longer zone than one migrating in the opposite direction (see Figure 8.1 b). When the conductivity of the sample solution is equal to that of the BGE, the length of the injected zone of an analyte i is:

$$L_{\mathrm{inj},i} = (\mu_{\mathrm{eo}} \pm \mu_i) \cdot \frac{V_{\mathrm{inj}}}{L} \cdot t_{\mathrm{inj}} \qquad (8.3)$$

where V_{inj} is the voltage applied during electrokinetic injection. Since the sample plug is not spread out parabolic as with hydrodynamic injection, the contribution of the injection to the final zone widths is relatively smaller. The standard deviation of a plug-like zone is approximately 0.3 times its length. This implies that with electrokinetic injection slightly higher injection volumes can be tolerated compared to hydrodynamic injection.

0009-5893/00   S-44-08 $ 03.00/0

With this electrokinetic method, the amount of an analyte ion introduced is dependent on its mobility. When two different analyte ions are present in the sample in the same concentration, different amounts of these ions will be injected. In practice this is not a problem, since this effect is dealt with during calibration. However, the amount injected also depends on the conductivity of the sample solution [3] (see also the next paragraph). This may be an easily overlooked source of irreproducibility when the salt concentration of samples is not controlled, for instance with samples of biological origin.

8.2 Matrix effects on injection

In this paragraph the influence on peak width and peak areas of the sample matrix, and in particular of the conductivity of the sample compared to that of the separation buffer, will be discussed. To simplify the discussion it will be assumed:

- that the concentration of the analyte is low, so that its influence on the conductivity of the sample solution (γ_{smp}) can be neglected;
- that both the background electrolyte and the sample matrix consist of a simple salt B^+C^-, with concentration c_{BGE} and c_{smp}, respectively;
- that the sample can be introduced as a sharp zone with length L_{inj}; with hydrodynamic injection this is actually not the case.

In the first stage of the separation part of the analyte ions are still in the sample zone with deviating conductivity, and consequently deviating field strength. According to the Kohlrausch rules, this deviating conductivity zone will stay intact during electrophoresis; it is a so-called stagnant zone moving with the electroosmotic flow (see paragraph 7.2). Considering that the current density j is constant in the system, and that $j = E.\gamma$, the field strength in the sample zone can be written as:

$$E_{smp} = \frac{\gamma_{BGE}}{\gamma_{smp}} \cdot E_{BGE} = \frac{c_{BGE}}{c_{smp}} \cdot E_{BGE} \qquad (8.4)$$

Now consider the period from the moment when the high voltage is switched on to the time that the last of the analyte ions have left the original sample zone (Figure 8.2). Since the capillary is almost completely filled with the BGE, the field strength in the BGE may be approximated as V_{appl}/L. The ions initially at the front of the zone migrate all of this period in the BGE, with a velocity of:

$$v_i = \mu_i \cdot E_{BGE} = \mu_i \cdot V_{appl} / L$$

The ions starting at the backside of the sample zone are in the sample solution during this period, migrating with a velocity:

$$v_i = \mu_i \cdot E_{smp} = \mu_i \cdot c_{BGE} / c_{smp} \cdot V_{appl} / L$$

The last of the analyte ions will have left the zone with deviating conductivity after a certain time t_{init}, which is

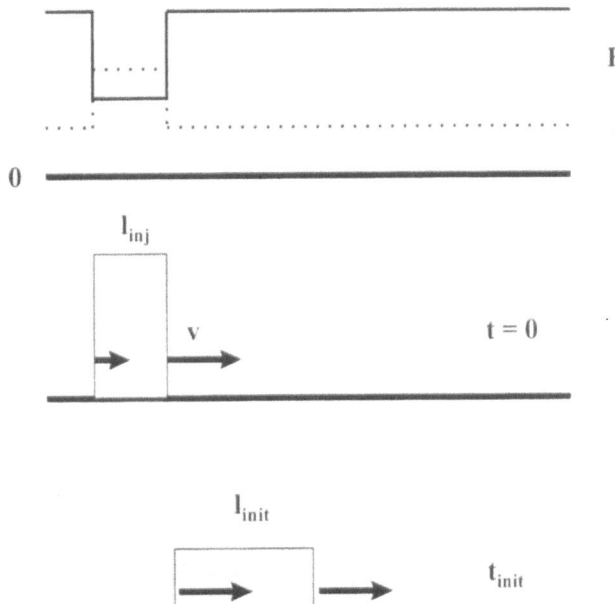

Figure 8.2
Broadening of the sample zone in the initial stage of the separation by caused by a high sample conductivity.

given by the length of this zone and the velocity above. Considering the actual separation to start after this initial time period, the zone width L_{init} is given by:

$$L_{init} = \frac{c_{smp}}{c_{BGE}} \cdot L_{inj} \qquad (8.5)$$

This means that when the salt concentration of the sample matrix is much larger than that of the BGE, the contribution to the zone broadening may be much larger than expected. To find the maximum sample amount that can be introduced before resolution is lost, the "initial" zone length as given in Equation 8.5 should be substituted in Equation 8.1.

Similar zone broadening effects are possible for weakly acidic or basic analytes, when there is a mismatch between the pH of the sample and that of the BGE. When the effective electrophoretic mobility of an analyte compound in the sample solution is much lower than in the BGE, its zone is again spread out during the initial stage of the separation. A pH mismatch can even lead to peak-splitting; a single (amphoteric) analyte compound may produce two or three different peaks in the electropherogram, representing the anionic, neutral and cationic species in the original sample matrix [4,5].

With pressure-driven sample introduction the sample matrix only influences the zone widths. When the effect is not excessive compared to other sources of zone-broadening, it may not even be observed. However, with electrokinetic sample introduction the peak areas are also affected by the sample matrix composition. With electrokinetic injection, both osmosis and migration play a role in the introduction of the sample. Since the

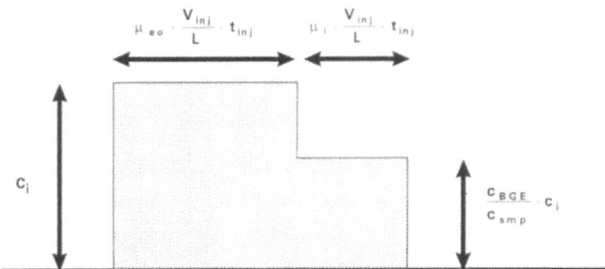

Figure 8.3
Concentration profile of a (cationic) analyte after electrokinetic injection.

capillary is still almost completely filled with the BGE, the osmotic flow is virtually not influenced by the sample conductivity. The part of the sample introduced by osmosis is not changed. However, the "extra" part introduced by electrophoretic effects during the injection depends on the conductivity of the sample. The total amount of analyte ions i injected is given by:

$$n_i = \frac{1}{4}\,\pi d_c^2 \cdot \left(\mu_{eo} \pm \frac{c_{BGE}}{c_{smp}} \cdot \mu_i \right) \cdot \frac{V_{inj}}{L} \cdot t_{inj} \cdot c_{i,smp} \quad (8.6)$$

In Figure 8.3 this equation is illustrated. When the sample conductivity is not regulated, i.e., large variations in γ_{smp} may occur from one sample to another, the sensitivity (peak area vs. sample concentration) will be irreproducible.

The sample conductivity may also influence the electrophoretic current and peak migration times. After (hydrodynamic) injection a low-conductivity sample plug can cause a high contribution to the total electric resistance of the solution in the capillary. The current flowing during electrophoresis will be lower than expected for the particular BGE. A disproportionate large part of the applied voltage will be over the sample plug; the electric field in the remainder of the capillary will be lower than expected, so that the EOF and the ionic migration is slow. Since the low-conductivity injection zone is quasi-stagnant, the low current and slow migration may last for a considerable time; a breakdown of the high resistance zone depends on axial diffusion of the BGE salt into the zone. Often the low current and field strength in the BGE continue until the stagnant zone leaves the capillary. Some time after the appearance of the EOF peak (depending on the length of the capillary behind the detection window) a stepwise increase of the current can then be observed.

8.3 Sample stacking

In the previous paragraph it was shown that when the sample conductivity is higher than that of the BGE, the sample zone is smeared out during the first part of the separation. The opposite may also occur; sample zones are compressed when the sample matrix has a lower conductivity than the BGE. This focusing effect may be

utilised to increase the loadability of the CE system; with samples having a low salt content, the sample volume can be much larger before the zone-broadening becomes too strong. This technique of preconcentration is called sample stacking (when the injection is performed hydrodynamically) or field-amplified injection (with electrokinetic injection) [6].

For analyte ions migrating in the same direction as the osmotic flow (usually cations), electrokinetic injection can be combined with simultaneous sample preconcentration. When the sample conductivity is low enough, e.g., when the sample matrix is pure water, the normal injection time can be applied, while still a large amount of sample cations are introduced into the capillary by the field-amplification effect (see Equation 8.6). Stacking (zone sharpening) occurs automatically at the interface between the sample solution and the BGE. It is often recommended first to introduce a small amount of water (or another blank low-conductivity solution) to take care of possible disturbances of the stacking process on the border between the BGE and the sample [7]. In Figure 8.4 the field-amplification effect is illustrated [8]. Electropherograms are compared obtained with a sample of PTH-amino acids dissolved in BGE and in pure water. From pure water, the positively charged analytes are strongly enriched during injection.

Field amplification can also be used for the enrichment of analyte ions migrating in the direction opposite to that of the EOF, when it is combined with the polarity switching technique as shown schematically in Figure 8.5. First, a plug of water is injected. Then, with the polarity of the high voltage in the normal direction, positive ions are selectively extracted from a low-conductivity sample solution. In the next step, the voltage polarity is reversed. Negative ions are drawn into the water plug. To keep part of the positive ions inside the capillary, the time of application of the reversed polarity should be shorter than the first injection time. Finally, the sample vial is replaced by the buffer vial and negative and positive ions are stacked on the two sides of the water plug.

A disadvantage of the (electrokinetic) field-amplification method is that the amount of sample ions introduced into the capillary strongly depends on the actual sample conductivity. When this conductivity can not be guaranteed, results will be unreliable. Therefore, hydrodynamic stacking methods are more often applied. Here, a long plug of a low-conductivity sample solution is introduced with a pressure difference. Due to the difference in conductivity, analyte zones are sharpened after the application of the electric field, in the first stage of the separation. When the conductivity of a particular sample is not as low as expected, the resulting peak resolution will be affected, but the peak areas are still reliable.

The amount of sample that can be injected, before the efficiency of the separation starts to deteriorate, is often more than 100-fold higher than without stacking. The

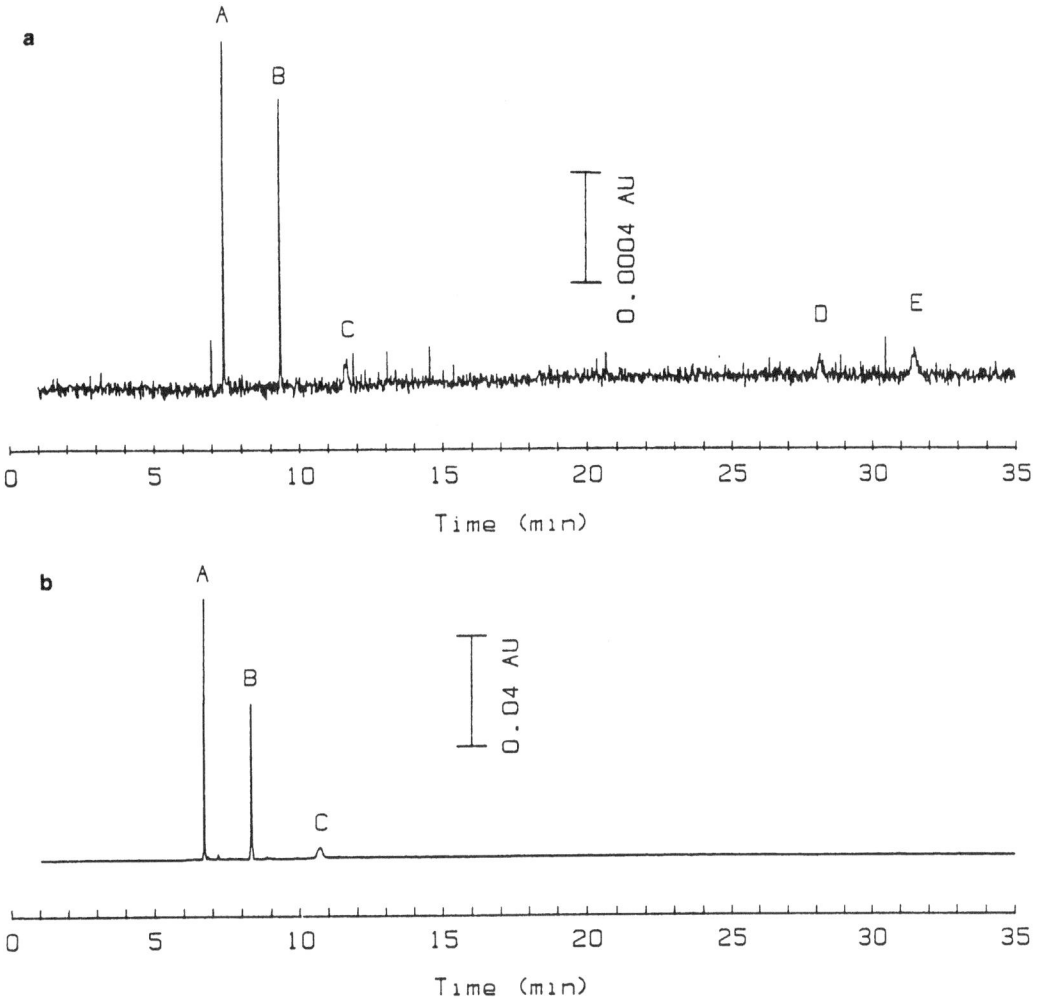

Figure 8.4

Influence of the sample matrix on electrokinetic injection. Electropherograms of a sample of basic (A,B) and acidic (D,E) PTH-amino acids dissolved in the BGE (a) or in water (b). Reproduced from ref. [8].

enrichment factor that can be realised by stacking, without compromising the separation, depends in the first place on the upper limit of the sample conductivity that can be guaranteed. However, there are also some other limitations. In the first place the Joule heating in the sample plug should be considered. When the conductivity of the sample matrix is much lower than that of the BGE, a relatively large part of the applied high voltage is over this sample plug, so that the field strength in this part of the capillary is higher than average. This may result in excessive heat generation in the sample solution, causing degassing and/or boiling. This excessive heat generation is especially strong when the low-conductivity sample plug is relatively short, i.e., when the stacking possibilities are not fully exploited.

Another limitation for sample stacking is related to differences in electroosmotic velocities occurring between the part of the capillary with the sample solution and the remainder of the capillary, containing the BGE. In the low-conductivity sample plug the field strength is higher than in the rest of the capillary. Moreover, the electroosmotic mobility is usually increasing with de-

creasing salt concentration. Both effects cause the electroosmotic velocity in the sample plug to be higher than in the BGE. Because of the "principle of conservation of volume", parabolic flows are superimposed on the plug-like osmotic flows in both parts of the capillary. As is shown in Figure 8.6, the resulting EOF is a weighted average of the intrinsic flows in the two capillary parts [9]. The superimposed parabolic flows, however, are a source of zone broadening, occurring not only in the sample plug but also in the BGE during the actual separation [10]. Especially when the difference in electroosmotic velocity is large, and when the sample zone is relatively long, the plate numbers obtained will be decreased. Manipulation of the electroosmotic mobility, either that in the sample plug or in the BGE, to create a better match, may help to decrease the peak broadening and to increase the sample volume that can be loaded [11].

When the analytes are weak cations or anions, sample stacking can also be realised by pH manipulation. When the pH of the sample is different than that of the BGE, the effective mobilities of the analyte ions can also be

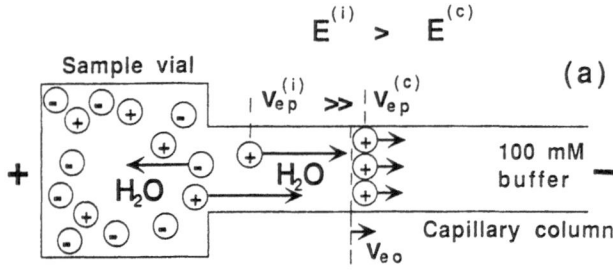

$E^{(i)} > E^{(c)}$

(a)

Sample vial

$|V_{ep}^{(i)}| \gg |V_{ep}^{(c)}|$

H_2O H_2O

100 mM buffer

Capillary column

V_{eo}

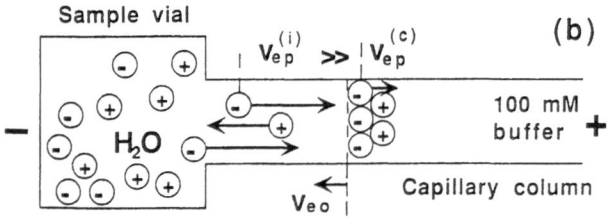

Sample vial

(b)

$|V_{ep}^{(i)}| \gg |V_{ep}^{(c)}|$

H_2O

100 mM buffer

Capillary column

V_{eo}

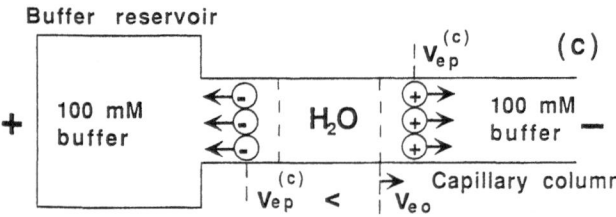

Buffer reservoir

(c)

$|V_{ep}^{(c)}|$

100 mM buffer H_2O 100 mM buffer

Capillary column

$|V_{ep}^{(c)}| < |V_{eo}|$

Figure 8.5
Diagram showing the procedure for field-amplified polarity-switching injection. Reproduced from ref. [8].

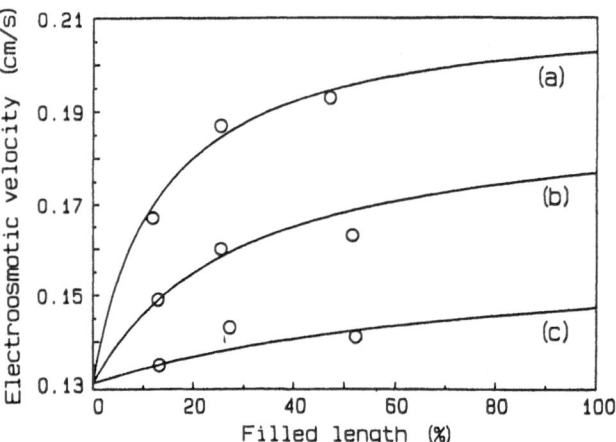

Figure 8.6
Influence of the length of the injected low-conductivity sample plug on the resulting EOF in field-amplified injection. Reproduced from ref. [9].

different. The pH of the sample and the BGE should be chosen such that the mobilities in the sample are (much) higher than in the BGE. Again, a large sample plug can then be introduced by pressure, which is compressed during the first stage of the separation.

Discontinuous buffer systems have been used for the stacking of samples that do not have a low conductivity

[12,13]. The stacking in these cases is based on iso-tachophoretic principles. In isotachophoresis (ITP) a sample solution is sandwiched between a leading and a terminating electrolyte solution [14]. The leading electrolyte contains a co-ion that has a higher mobility than any of the analyte ions; the terminating electrolyte one that is slower than the analyte ions. Since during electrophoresis the analyte ions in such a system can not overhaul the leading ion, and the terminating ion not the analyte ions, sharp (pure) zones of analytes are formed, that migrate (all with the same speed) between the two electrolytes in an order according to the mobilities of the ions. The concentrations of the pure analyte zones in ITP are determined by the composition of the leading and terminating electrolytes (following the Kohlrausch regulation), and not by the original concentrations in the sample solution. Therefore, sample enrichment or stacking is a common phenomenon in ITP.

The practical application of (on-capillary) ITP-CE [15,16] for the stacking of cations is illustrated in Figure 8.7. The BGE for the CE separation serves as the leading electrolyte (L); therefore, it has been chosen such that its co-ion has a higher mobility than any of the analyte ions. First the capillary is flushed with the BGE/L solution. Then, a large plug of the sample solution is injected hydrodynamically. Next, the capillary inlet is placed in a vial containing the terminating electrolyte (T), and an electric field is applied for isotachophoretic focusing of the analyte zones. A hydrodynamic back pressure is applied on the end side of the capillary to prevent too much of the terminating electrolyte to enter the capillary by electroosmosis. When focusing is complete, the remainder of the terminating solution is removed from the capillary by pressure, and the capillary inlet is placed in a vial with the BGE/L solution. Then, the actual zone-electrophoretic separation of the stacked analytes can start. The zone-electrophoretic separation can also be carried out with the terminating electrolyte as the BGE by using a slightly different ITP procedure, including polarity reversal of the applied voltage. Similar schemes have been developed for anionic analytes.

8.4 Coupled ITP-CE

As discussed above, ITP-CE can be performed in one capillary. However, with a two-capillary system an even higher sample enrichment can be realised [17]. Figure 8.8 shows schematically the set-up that can be used for such a coupled-capillary system [18]. A commercial ITP instrument was modified for this purpose. A wide capillary (500 μm I.D.) was used for the ITP part of the set-up, so that large volumes of sample can be loaded. Closely behind the detector that is used to monitor the passage of the ITP-zones of interest, a narrow (50 μm I.D.) CE capillary has been inserted. A grounded electrode in the leading electrolyte vial served as cathode both in ITP and in CE; a single high voltage source, switched between the terminating electrolyte vial and the end vial for the CE capillary, was used for both-

stages of the separation. After the ITP focusing, a part of the volume containing the enriched analyte zones is injected electrokinetically into the CE capillary. Compared to the single-injection mode, detection limits could be improved by more than two orders of magnitude.

The limit for the enrichment factor that can be reached in coupled ITP-CE is not related to the ITP process itself. The analyte concentration that is present in a focused zone after ITP is not dependent on its original concentration in the sample solution. However, the total volume of the zones focused between the leading and terminating electrolytes may become a problem. This volume will increase proportionally with the sample volume introduced into the ITP channel. Apart from the analytes this volume may also contain non-relevant ionic constituents of the sample solution. With increasing total zone volume it will become increasingly difficult to sample a representative portion of it into the separation capillary.

As in single-capillary ITP-CE, the zone-electrophoretic separation can be carried out in the leading electrolyte as BGE, or in the terminating electrolyte. In both cases it is important to avoid the introduction of the other ITP electrolyte into the CE capillary as much as possible. With some of the other electrolyte in the CE capillary, the ITP process continues partially for some time during the CZE-stage of the procedure. This leads to destacking of the analyte zones and subsequently to broadened and shifted analyte peaks in the electropherogram [19].

8.5 Chromatographic preconcentration

Preconcentration of analytes by stacking techniques or by ITP is attractive because it can be performed without special instrumental modifications. However, the enrichment that can be obtained with these techniques is limited by the possible presence of non-relevant ionic compounds in the samples to be preconcentrated. Another disadvantage of these analyte preconcentration methods is that little or no selectivity is added to the method by them. With stacking all ionic compounds in the sample are concentrated; with (one-capillary) ITP a broad range of analytes is selected with mobilities between those of the leading and terminating ions. Therefore, much research effort has been put in the development of other preconcentration techniques for CE. Techniques based on chromatographic principles are especially attractive because they can provide extra selectivity to the analysis.

Of course it is well possible to combine CE off-line with one of the sample preparation techniques that are routinely applied with HPLC, such as automated liquid-liquid extraction (LLE) or solid-phase extraction (SPE). An advantage of working off-line is that it is generally easy to design a method yielding a final solution that is compatible with CE. Often it is possible to concentrate the analytes in a low-conductivity solution, so that the analyte enrichment can be further enhanced by applying

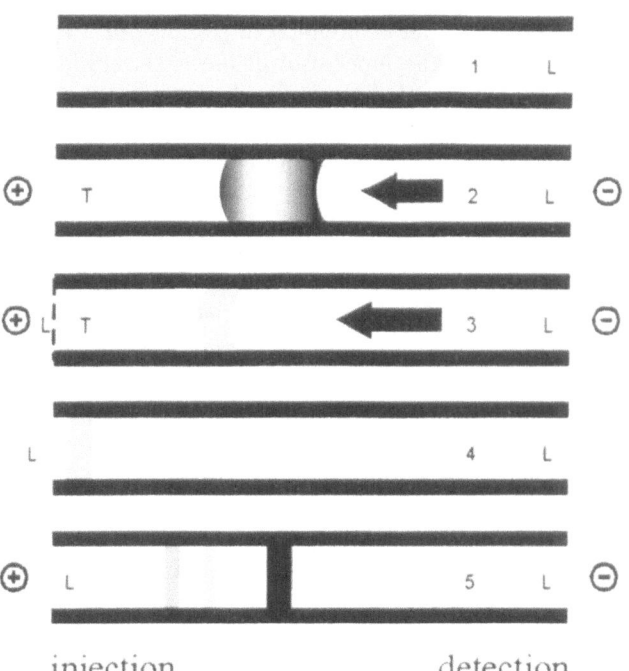

Figure 8.7
Diagram showing the procedure of on-capillary ITP-CE for cations with hydrodynamic back pressure. Reproduced from ref. [15].

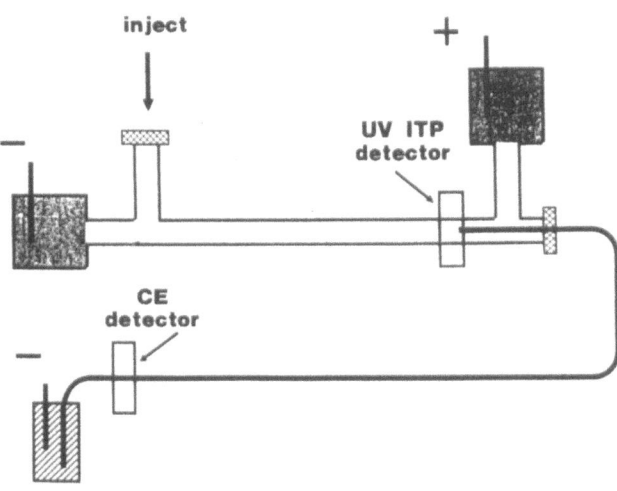

Figure 8.8
Scheme of the set-up for two-capillary ITP-CE. Reproduced from ref. [18].

a sample stacking technique. The only negative aspect of using a standard off-line sample-prep apparatus that can be mentioned is that these instruments usually work on a volume scale that is overdone for CE. However, recently an automated SPE instrument has been marketed that works with microliter volumes.

Several research groups have worked on the development of on-line chromatographic preconcentration techniques for CE. One of the approaches is to use two fused silica capillaries coupled in series with a PTFE sleeve [20,21]. The inner wall of the first capillary is modified with a retentive layer; the selectivity of the method can of course be determined by the choice of this retentive layer. The second capillary is a normal fused silica capillary used for the separation. In the pre-concentration step of the procedure the (binding) sample solution is fed to the capillary system, either by pressure or electrokinetically. The sample matrix solution should be chosen such that the analytes of interest show a high affinity for the retentive layer on the wall. In the next step, a debinding solution is fed to the system. The analytes are collected in a small volume at the front of this debinding solution. This solution also acts as the BGE for the separation. A point of concern with this technique is the low capacity of the retentive layer on the inner wall of the preconcentration capillary. It has been shown that the loadability can be improved by using a capillary with a roughened inner surface [19].

An alternative approach for on-line SPE-CE is the packing of a segment of the capillary system with an appropriate stationary phase material [22,23,24]. Generally, a separate piece of capillary is packed with the stationary phase, often a C8 or C18 modified silica material. The phase has to be sandwiched between to frits or supports, to prevent it from blocking the separation capillary. The miniature precolumn is connected to an open capillary by means of PTFE sleeves (see Figure 8.9). In the first step of the procedure with this system, the sample solution is loaded onto the precolumn by means of pressure. Next, the precolumn is washed with an appropriate solution that can serve as the BGE for the CZE separation. Then the adsorbed analytes can be desorbed by introducing a relatively small volume of a desorbing solution. Finally, the inlet of the capillary is placed in a vial with BGE and the separation can be started.

Although it has been shown by different researchers that on-line SPE-CE can provide impressive enrichment factors, its application is still far from routine work. There are still two major problems to be solved. The first problem is the compatibility of the desorbing solution with the CE separation. To desorb the analytes from the (hydrophobic) stationary phase in most cases a solution with a high organic modifier content is used. It is often observed that the presence of a relatively long plug of this solution in the capillary disturbs the electrophoretic separation. For every application a solution with enough desorbing strength and suitable properties during the separation has to be found. A more fundamental problem is the disturbance of the electroosmotic flow by the presence of the packed part in the capillary system. Even when this part is relatively short, it has a strong influence on the resulting EOF because of its high flow resistance [25]. In general, the electroosmotic velocity in the packed bed will not match that in the open tube, so that parabolic flows deteriorate the separation efficiency. Moreover, it has been observed that the EOF can be strongly influenced by the adsorption of sample ions

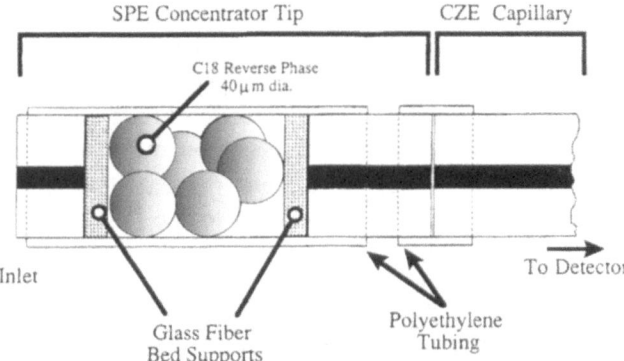

Figure 8.9
Schematic representation of a solid-phase concentrator and its attachment to the CE capillary for on-line SPE-CE. Reproduced form ref. [26].

on the stationary phase during sample loading [26]. This will make migration times strongly irreproducible.

8.6 Hyphenated techniques

The low sample volume required and its high speed make CE a suitable candidate as the second technique for two-dimensional separations. The main technical problem in the hyphenation of CE to another separation technique is related to the high voltage applied. In ordinary CE the detection side of the capillary is usually grounded for safety reasons; this is the side most likely to make contact with other instrumentation or personnel. With CE as a second separation technique, either the grounding has to be on the injection side, or the point of hyphenation should be thoroughly insulated.

Comprehensive 2D separation systems, coupling column chromatography to fast CE, have been introduced by Jorgenson c.s. [27,28]. In the first version of their system a (grounded) computer-controlled six-port valve was connected to the outlet of the LC column. By means of this valve the inlet of the CE capillary could be switched between the column effluent and a flowing stream of blank BGE (see Figure 8.10). The speed of the switching valve did not permit very short sampling times. Therefore, the CE voltage was lowered during the

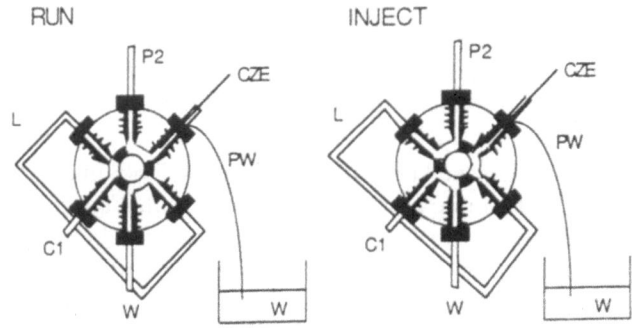

Figure 8.10
Six-port valve used for the coupling of LC and CE. C1: HPLC column outlet; P2: pump with BGE; PW: paper wick used as drain for column effluent. Reproduced from ref. [27].

electrokinetic sampling of the column effluent, to avoid volume overloading of the CE system. By using a short separation capillary (6.5 cm to the detector) the CE run time could be kept short, so that the LC column effluent could be sampled every minute. In Figure 8.11 a two-dimensional "chromatoelectropherogram" is shown of an ovalbumin tryptic digest obtained with this system. The lack of a specific diagonal pattern in the plots clearly illustrates the orthogonality of the two separation systems.

To profit optimally from a 2D separation system, in terms of resolution and selectivity, the sampling rate for the second dimension should be as high as possible. In later work by the same group techniques and devices have been developed that permit sampling rates in the order of seconds [29,30].

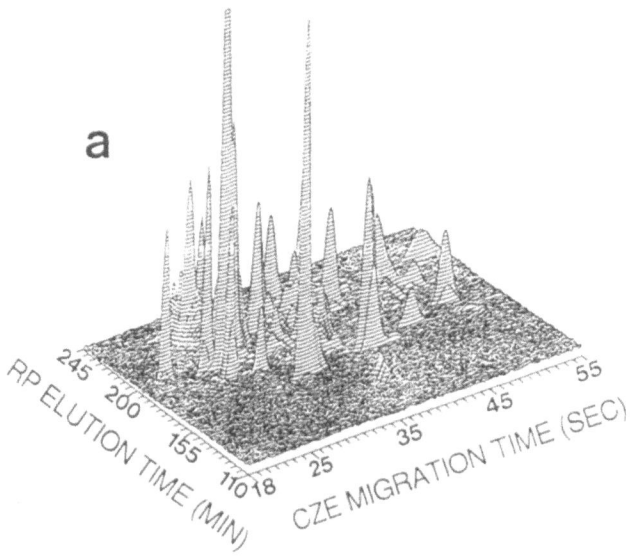

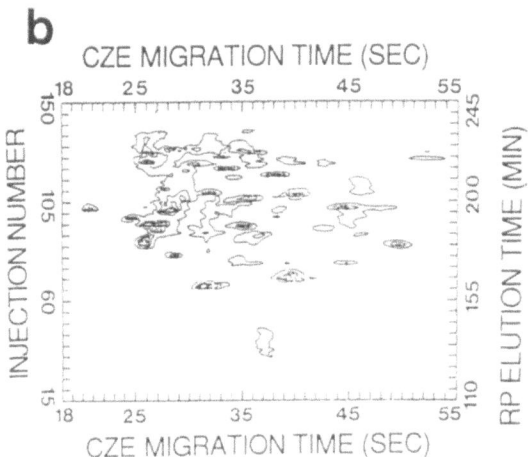

Figure 8.11
Comprehensive two-dimensional separation of a ovalbumin tryptic digest by LC-CE. (a) 2D plot, (b) contour plot. Reproduced from ref. [27].

References

[1] K.D. Altria and J. Bestford, J. Cap. Electrophoresis, 3 (1996) 13.
[2] K.D. Altria and H. Fabre, Chromatographia, 40 (1995) 313.
[3] X. Huang, M.J. Gordon and R.N. Zare, Anal. Chem., 60 (1988) 375.
[4] S.V. Ermakov, M.Y. Zhukov, L. Capelli and P.G. Righetti, Anal. Chem., 66 (1994) 4034.
[5] S.V. Ermakov, M.Y. Zhukov, L. Capelli and P.G. Righetti, Anal. Chem., 67 (1995) 2957.
[6] R.-L. Chien and D.S. Burgi, Anal. Chem., 64 (1992) 489A.
[7] R.L. Chien and D.S. Burgi, J. Chromatogr., 559 (1991) 141.
[8] R.L. Chien and D.S. Burgi, J. Chromatogr., 559 (1991) 153.
[9] R.-L. Chien and J.C. Helmer, Anal. Chem., 63 (1991) 1354.
[10] W.Th. Kok, Anal. Chem., 65 (1993) 1853.
[11] D.S. Burgi, Anal. Chem., 65 (1993) 3726.
[12] S. Hjerten, K. Elenbring, F. Kilar, J. Liao, A.J.C. Chen, C.J. Siebert and M. Zhu, J. Chromatogr., 403 (1987) 47.
[13] C. Schwer and F. Lottspeich, J. Chromatogr., 623 (1992) 345.
[14] F.M. Everaerts, J.L. Beckers and Th.P.E.M. Verheggen, "Iso-tachophoresis: Theory, Instrumentation and Applications", Elsevier, Amsterdam, 1976.
[15] N.J. Reinhoud, U.R. Tjaden and J. van der Greef, J. Chromatogr., 641 (1993) 155.
[16] N.J. Reinhoud, U.R. Tjaden and J. van der Greef, J. Chromatogr., 653 (1993) 303.
[17] D. Kamiansky and J. Marak, J. Chromatogr., 498 (1990) 191.
[18] D.S. Stegehuis, H. Irth, U.J. Tjaden and J. van der Greef, J. Chromatogr., 538 (1991) 393.
[19] L. Krivankova, P. Gebauer and P. Bocek, J. Chromatogr. A, 716 (1995) 35.
[20] J. Cai and Z. El Rassi, J. Liq. Chromatogr., 15 (1992) 1179.
[21] J. Cai and Z. El Rassi, J. Liq. Chromatogr., 16 (1993) 2007.
[22] N.A. Guzman, M.A. Trebilcock and J.P. Advis, J. Liq. Chromatogr., 14 (1991) 997.
[23] M.E. Swartz and M. Merion, J. Chromatogr., 632 (1993) 209.
[24] C.L. Ng, C.P. Ong, H.K. Lee and S.F.Y. Li, J. Chrom. Sci., 32 (1994) 121.
[25] A.J. Tomlinson, L.M. Benson, R.P. Oda, W.D. Braddock, M.A.Strausbauch, P.J.Wettstein and S.Naylor, J. High Resolut. Chrom., 17 (1994) 669.
[26] M.A. Strausbauch, J.P. Landers and P.J. Wettstein, Anal. Chem., 68 (1996) 306.
[27] M.M. Bushey and J.W. Jorgenson, Anal. Chem., 62 (1990) 978.
[28] M.M. Bushey and J.W. Jorgenson, J. Microcolumn Sep., 2 (1990) 293.
[29] A.W. Moore and J.W. Jorgenson, Anal. Chem., 67 (1995) 3448.
[30] A.W. Moore and J.W. Jorgenson, Anal. Chem., 67 (1995) 3456.

9 UV-Absorption Detection

9.1 Introduction

The measurement of the absorption of light in the UV or visible region is by far the most popular detection principle in CE. All suppliers sell instruments with a single- and/or multi-wavelength absorption detector, while other detector options, such as a lamp- or laser-induced fluorescence detector or a conductivity detector, are commercially available from only a few manufacturers. Probably over 95 % of the routine applications of CE is carried out with a UV detector. The reasons for the popularity of this detector are obvious:

- The detection can be performed on-column, i.e., directly on the capillary, and without the need for physical contact between the separation channel and the measurement system. By working on-column, the detector zone broadening can easily be kept negligible. Because of the absence of physical contact with the separation medium, there are no problems with interference of the high voltage used for the separation with the sensitive, vulnerable electronic parts of the detection unit.

- Fused silica, of which the capillaries used in CE are commonly made, is a quartz-like material which is transparent for light with wavelengths down to 190 nm. Therefore, the capillary wall itself makes a perfect optical cell. In fact, the use of alternative wall materials, with for instance better surface properties in respect to the adsorption of certain analytes, is often precluded because they are not transparent for UV light.

- In the decade before the emergence of CE, important progress had been made in the instrumental development of sensitive, low-noise UV detectors for HPLC. It appeared that only some small modifications in the optical system were necessary for the use of these detectors in CE. Therefore, variable wavelength UV detectors for CE are usually equipped with the familiar deuterium lamp for the UV region or a tungsten lamp for the visible region, a grating for wavelength selection, a beam splitter and two solid-state photodiodes for the monitoring of the light intensities of the sample and reference beams. Multi-wavelength detectors (with a diode array or a fast-scanning grating) are also not fundamentally different from those used in HPLC.

Still, for those who started with CE as an alternative for HPLC, the performance of UV detection was somewhat disappointing. The reason for this is immediately clear from the Lambert-Beer law, the fundamental principle of absorption detection. The signal produced by the detector, the absorbance A, is given by:

$$A = \varepsilon \cdot c \cdot L_{\text{eff}} \tag{9.1}$$

where ε is the molar absorptivity of the analyte, c its concentration and L_{eff} the effective length of the light

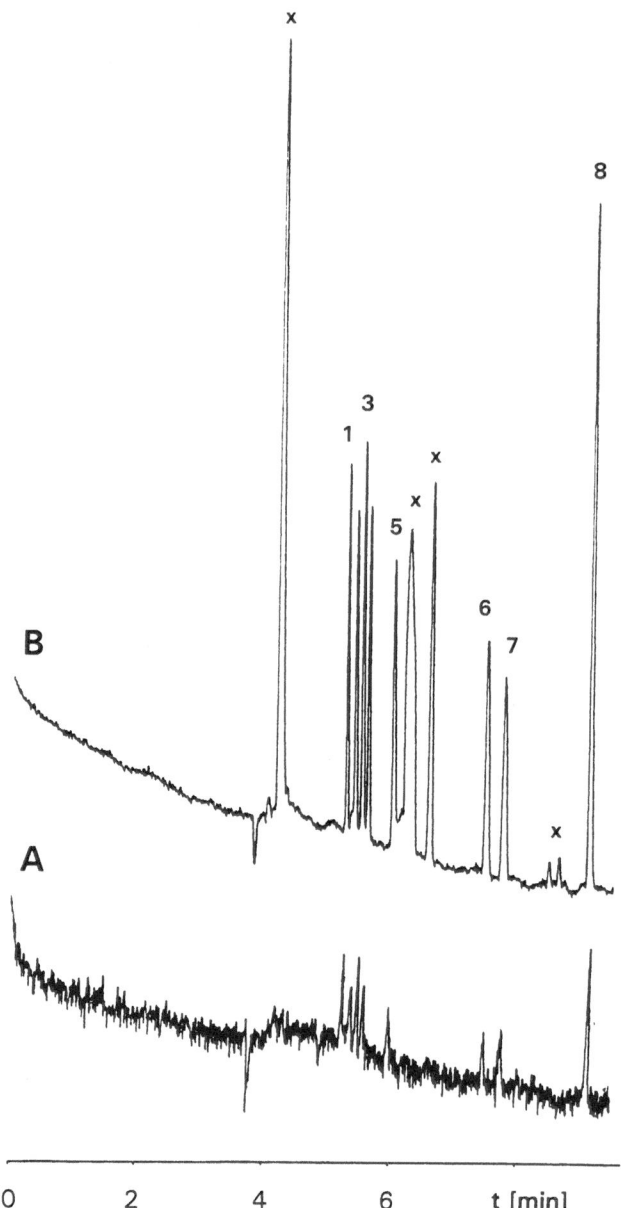

Figure 9.1

Sensitivity versus selectivity by wavelength selection in CE with UV detection. Separation of OPA-derivatized amino acids with detection at (A) 335 nm " and (B) 220 nm. Sample concentration 10^{-5} mol L^{-1}; both electropherograms are recorded with a sensitivity of 0.005 AU full scale. Peaks: 1 = SER; 2 = GLY; 3 = ALA; 4 = HIS; 5 = TYR; 6 = GLU; 7 = ASP; 8 = LYS. The peaks denoted with an × are impurities from the derivatization reagent.

path through the solution. The sensitivity (A/c) for a particular compound is then proportional to L_{eff}. When the light absorbance is measured in a direction perpendicular to the capillary axis, the light path length L_{eff} can be no more than the capillary diameter. For a typical well-absorbing compound ε is in the order of 5 to 10,000 mol^{-1} L cm^{-1}. With a noise level of 1 to 5 × 10^{-5} absorbance units, the detection limit for such a compound, when using a 50 μm ID capillary, will be in the order of 1 μmol L^{-1} at the detection window, and higher than that in the sample.

0009-5893/00   S-52-07 $ 03.00/0   © 2000 Friedr. Vieweg & Sohn Verlagsgesellschaft mbH

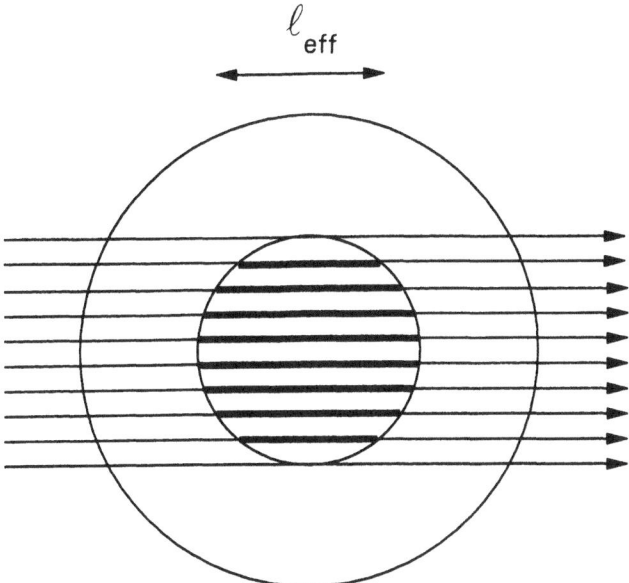

ℓ_{eff}

Figure 9.2
Variation of the light path length over the diameter of the capillary in on-column UV detection. The thicker line parts show the actual local path lengths. Refraction and reflection phenomena are neglected for simplicity.

With UV absorption measurements, limits of detection in CE are approximately two orders of magnitude higher than in HPLC. However, in some cases part of this disadvantage can be compensated for by choosing a different wavelength for detection. The selectivity of the CE separation is often so high that a less selective wavelength can be used. Most organic compounds have a very high molar absorptivity near 200 nm. An example of this is shown in Figure 9.1. In a study on the performance of different detection modes in CE, OPA-derivatives of amino acids were separated. With a UV detector operated at a selective wavelength (335 nm), detection limits were in the order of 10^{-5} mol L^{-1}. At 220 nm on the other hand, the sensitivity was higher and the noise level decreased. Detection limits in the order of 10^{-6} mol L^{-1} could be obtained, at the expense of a lower selectivity; a number of peaks are visible in the electropherogram from impurities in the derivatization reagent.

9.2 The sensitivity and linearity of UV detection

When UV detection is performed perpendicular to the capillary axis, as it usually is, the optical cell has a cylindrical shape as is shown schematically in Figure 9.2. When this cell is illuminated by a parallel beam of light with a certain width (w_b), the length of the light path is varying over the width of the beam. The transmission of light through such a cylindrical cell has been treated theoretically by Hjerten [1] and by Bruin et al. [2]. The effective light path length (L_{eff}) depends not only on the capillary inside diameter d_c, but also on the ratio of the light beam width and d_c. Only when the beam

is very narrow, d_c can be substituted for L_{eff} in Equation 9.1 to predict the sensitivity of detection. When the beam of light is exactly as wide as the capillary inner diameter, the path length varies from 0 at the top and the bottom to d$_c$ in the middle of the capillary cross-section. The average value L_{eff} is then equal to 0.79 d_c.

When the light beam is wider than the capillary inner diameter, part of the light will even pass the capillary undisturbed by absorbance in the solution. This will decrease the sensitivity of detection. To understand this, one should realise that the signal produced by a UV detector is based on a comparison of the intensity I of the light beam passing the optical cell with a reference value I_0. The ratio of I and I_0 is translated electronically into the signal A:

$$A = -\log \frac{I}{I_0} \qquad (9.2)$$

In most single or variable wavelength detectors a second, undisturbed light beam collected at a second photodiode is used as reference. With diode-array detectors a part of the spectrum where no absorbance is expected can be used for reference. When now a part of the light passes the capillary undisturbed, because of insufficient focusing of the light beam, the "amount" of light absorbed by compounds in the solution is not affected. However, relative to the total intensity, the change of I by absorption will be smaller. The lower change of I/I_0 is then translated into a lower signal.

Alternatively, one can state that the effective (average) light path length is decreased by light missing the core of the capillary. With some simplifying assumptions [2], a formula can be derived relating L_{eff} with the width of the light beam w_b, which is valid when the light beam is wider than the capillary core:

$$L_{\text{eff}} = \frac{0.79 d_c^2}{w_b} \qquad (9.3)$$

This equation shows that the sensitivity of the detector, which is proportional to L_{eff}, may depend on the focusing of the light beam. For cuvette spectrophotometers and HPLC detectors, with the standard 10 mm path length, one may expect that instruments obtained from different suppliers have at least the same sensitivity. For CE detectors this is not necessarily true. Differences in sensitivity are not related to differences in lamp intensity or of the quality of the photodetectors used in the instruments; such differences are cancelled out because the signal is obtained from the ratio of two measurements (I/I_0). Especially when very narrow (e.g., 25 μm) capillaries are used, the sensitivity obtained with an instrument depends on the ability of the manufacturer to focus the light sufficiently.

The performance of different instruments in this respect can be compared and quantified by measuring the effective path length for, e.g., the standard 50 μm capillary. For this, one should do the following.

Table 9.1 Dependency of the effective light path length on the capillary inner diameter and the width of the illuminating beam. Experimental data taken from reference [2].

d_c [μm]	L_{eff} [μm]			
	$w_b = 50\,\mu m$	$w_b = 100\,\mu m$	$w_b = 150\,\mu m$	experimental
25	10	5	3	3
50	39	20	13	15
75	69	44	29	31
100	96	79	52	56

- Prepare a sample solution (in the BGE) of a compound with known or measured molar absorption ε at the wavelength to be used for the application at hand. Care should be taken that the concentration of the compound is low enough to be within the linear range of the sensitivity (see below); the value of $\varepsilon \cdot c$ should be below 0.1.

- Fill the capillary with the BGE. Flush a large plug of the sample solution through the capillary by applying pressure on the inlet of the capillary. Measure the absorbance at the appropriate wavelength at the plateau of the sample plug.

- From the absorption A as given by the instrument, L_{eff} can be calculated directly as:

$$L_{eff} = A / (\varepsilon \cdot c)$$

- The value of l_{eff} obtained may depend not only on the capillary inner diameter, but also on its outer diameter and the wavelength used.

Equation 9.3 also shows that with narrow capillaries (narrow in comparison to w_b), the sensitivity depends on the capillary cross-section (d_c^2) rather than on its diameter, as would be expected on basis of the Lambert-Beer law (Equation 7.1). The loss of sensitivity when one decreases the capillary diameter, for instance to reduce Joule heating effects, may be larger than assumed beforehand. In Table 9.1 calculated values of L_{eff} are given for various capillary diameters and beam widths. Also included in the table are experimental data obtained with a particular commercial instrument.

The discussion above on the effect of the quality of focusing of the light on the sensitivity, is only valid for low absorbances. However, there is also an effect on the linearity of the detector response. This is especially of importance for the indirect detection mode, where changes of the absorbance on a high background are measured (see Chapter II). Due to the cylindrical shape of the detection cell and the possible detection of light not passing the core of the capillary, the sensitivity is decreased for high absorbances. Qualitatively this can be explained as follows. With increasing concentration of the analyte, the intensity of the light transmitted through the centre of the capillary approaches zero. Normally, with a rectangular optical cell, this decreasing light intensity is translated into a signal linear with the absorbance ($-\log I/I_0$). With the capillary itself as

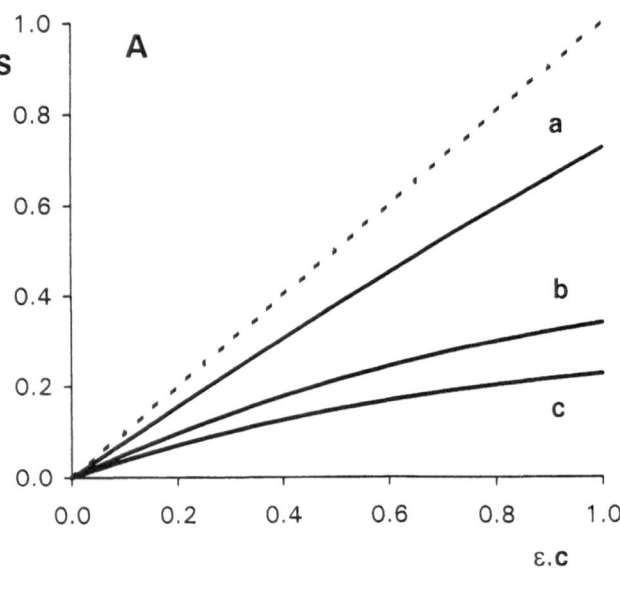

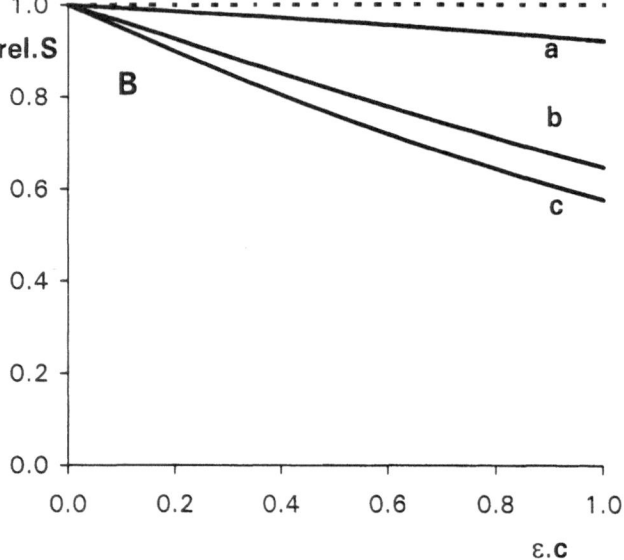

Figure 9.3
Influence of the quality of light focusing on the linearity of on-column UV detection. (A) Calculated calibration curves; (B) sensitivity of detection relative to that at infinite dilution. Light beam width w_b equal to (a) d_c; (b) 1.5 d_c; (c) 2 d_c. The dotted lines represent the ideal curves, with $w_b \ll d_c$.

optical cell, however, the relative importance of the light passing the solution with a shorter path length, or not passing the solution at all, becomes larger when the intensity of the light through the middle of the capillary becomes smaller by absorbance. In other words, the relative change of the total light intensity by an increase of the analyte concentration becomes smaller at higher concentrations.

To illustrate these effects, the theoretical relation between the signal and the molar absorbance and concentration of an analyte was evaluated by numerical approximation. The results are shown in Figure 9.3A for various ratios of the light beam width w_b and the capillary inner diameter d_c. The signals expected can be compared with the "ideal" case of a rectangular cell with an optical path length equal to d_c. At low absorbances, the figure shows the decreased sensitivity related to an insufficient focusing of the light, as discussed above. At higher values of εc, it can be seen that the nonlinearity of the detector response becomes more severe with increasing w_b/d_c ratio. This is shown more clearly in Figure 9.3B, where the sensitivity is plotted relative to that at low concentrations. Even when the light beam is not wider than the capillary inner diameter, a substantial decrease (approximately 8%) of the sensitivity is expected for values of εc around 1. It should be realised that for a reasonably well-absorbing compound, with e.g. an ε of 5000, this is already the case at concentrations in the order of 10^{-4} mol L^{-1}. The moral of all this is that the linearity of a CE method with on column UV detection should always be checked.

9.3 Optical arrangements for UV detection

In the previous paragraph the importance of the focusing of the light, to obtain an optimal sensitivity and linearity of response, has been discussed. It was shown that the light beam should preferably be focused to a width narrower than the inner diameter of the capillary. With the conventional HPLC detector such focusing is not realised, so that modifications or a renovation of the optical design appeared to be necessary.

In one of the earlier UV detectors for CE, reference and sample light beams were narrowed by 100 μm pinholes [3]. The disadvantage of such an arrangement is that the intensity of the light passing the pinholes is low, so that the noise level may be increased. This problem can be partly overcome by using an adjustable, rectangular slit, parallel to the capillary axis, for focusing [4]. In the axial direction the tolerance for the beam width is much larger than perpendicular to the capillary. Only in extreme cases, with very short capillaries and/or very high plate numbers, the detection window length should be limited to less than 1 mm (see paragraph 3.1).

In a number of detectors available on the market a sapphire ball is applied as a lens. Sapphire has a high refractive index ($n = 1.77$ for light in the visible region), so that a parallel beam of light directed on a ball is focused shortly behind it (see Figure 9.4A). The focal point with a 2 mm lens is at a distance of 100 to 150 μm, depending on the wavelength. With an appropriate holder (Figure 7.4B), the capillary can be fixed against the ball lens, so that the solution in its bore catches a large part of the light. The actual performance of the arrangement depends not only on the inner diameter but also on the wavelength and the outer diameter of the capillary. The photocell can be placed directly behind the capillary. With an extra slit stray light of deviating direction, reflected from the surfaces of the lens or the capillary, can be excluded from the photocell.

With a ball lens there is also focusing in the axial direction. As discussed above, this may be advantageous for the plate number in very demanding applications. In

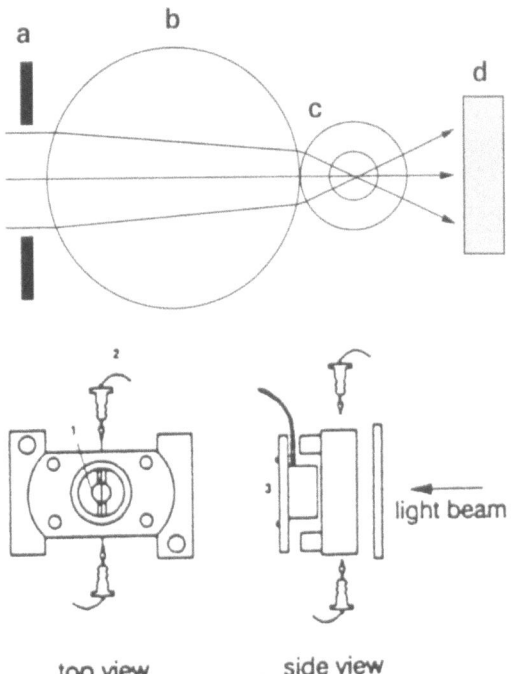

Figure 9.4
(A) Ray traces in an optical arrangement with a sapphire ball lens. For simplicity, reflected rays are omitted from the figure. A = aperture; B = ball lens; C = capillary; D = photodetector. (B) Schematic view of a capillary holder for a ball lens arrangement. 1 = aperture with ball lens; 2 = capillary; 3 = photodetector.

Figure 9.5 the performances of a rectangular slit and a ball lens arrangement can be compared in the separation of oligonucleotides (pdA$_{40\text{-}60}$) by capillary gel electrophoresis. With the ball lens, the resolution of the peaks is clearly improved.

In some detectors the light from the lamp is guided to the optical cell by use of optical fibres. An easier accessibility and better possibilities for the temperature control of the capillary close to the detection window are advantages of the application of fibre optics; transmission losses, especially for low wavelengths, can be a disadvantage.

9.4 Extended light-path detection

The obvious way to increase the sensitivity of UV detection in CE is by increasing the capillary inner diameter. Limits are set to this by the zone broadening and average temperature rise occurring with the increased Joule heat production in wider capillaries. However, there is nothing against using a capillary which is wider only locally at the detection window, so that the effective light path length is increased. Such so-called bubblecell capillaries have been brought on the market by Hewlett-Packard [5]. Capillaries with various expansion factors (the ratio of the diameter at the detection window to that of the rest of the capillary) are available. *In respect to Joule heating the bubble cell is even advantageous. At the detection position on the capillary, where it is always difficult to control the temperature, the field strength and with that the heat production is reduced, inversely proportional to the cross section of the capillary.*

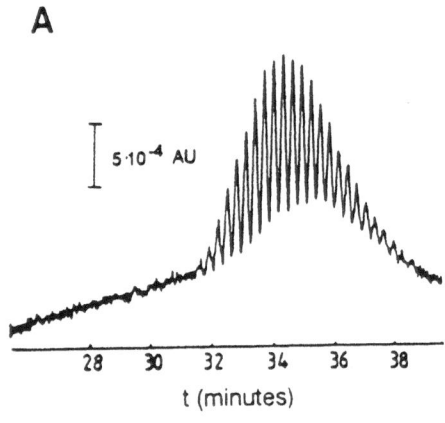

A

5·10⁻⁴ AU

t (minutes)

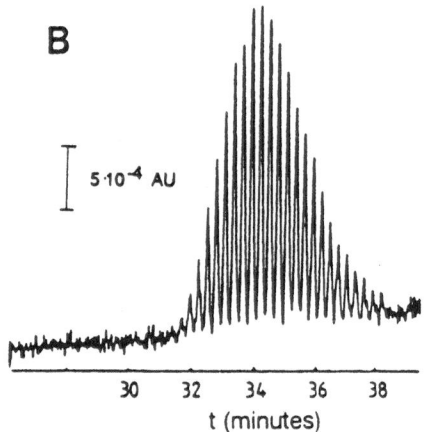

B

5·10⁻⁴ AU

t (minutes)

Figure 9.5
Influence of the beam width in the axial direction on peak resolution. Optical cell with (A) a rectangular slit, or (B) a ball lens system. Electropherograms of the separation of $p(dA)_{40-60}$ by capillary gel electrophoresis. Reproduced from reference [2].

It is easy to show that the radial expansion of an analyte zone in the bubble is accompanied with a sharpening of the zone in the axial direction: the volume of the zone is not affected [6]. Still, this does not improve the plate number, since the linear velocity of the zone is decreased correspondingly. On the other hand, it has also been shown [7] that changes of the capillary inner diameter do not cause much extra zone broadening when such changes are not too large or too abrupt.

The shortening of zone lengths in a bubble cell necessitates a stricter limitation of the light beam width in the axial direction. When in a conventional capillary a detection window width of 2 mm would be appropriate, in a bubble cell with an expansion factor of 3 a slit width of 200 μm would be required. Still, it can be imagined that it is easier to direct a high intensity of light through an aperture of $200 \times 150\,\mu$m that through one of $2\,\text{mm} \times 50\,\mu$m. Improvements of the limits of detection by a factor of 3 to 8 have been reported with bubble cells.

With some experience it is not difficult to fabricate a bubble-window in a capillary by glass-blowing procedures. Alternatively, one can just simply insert a short piece of a wider bore capillary between two pieces of the separation capillary. The wider detection window can be

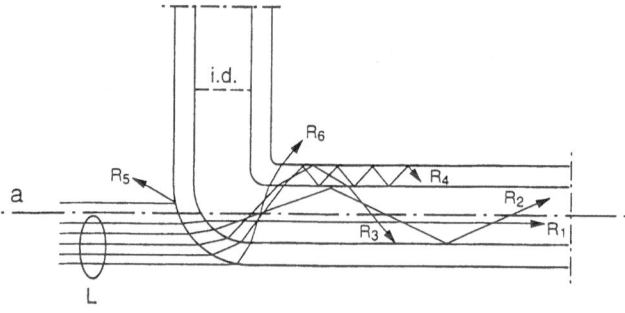

Figure 9.6
Ray traces for a Z-cell configuration. Rays R_1, R_2, and R_3 contribute to the signal; R_5 and R_6 are escaping by reflection and refraction, respectively; R_4 is trapped in the silica wall. Figure by courtesy of LC Packings (Amsterdam, Netherlands).

sealed with a resin. Good results have been obtained with this simple extended light path system [8].

Another approach to obtain a longer light path is by illuminating a part of the capillary in the axial direction [9,10]. To realise this, the fused silica capillary can be bent into a Z-shape and the transverse part used as the optical cell [11]. In the earlier attempts to apply a Z-cell in CE, high noise levels were obtained since the intensity of the light that could be directed through the cell was insufficient. Later, the light coupling was improved with a sapphire ball lens arrangement [12]. The light focused by the lens is projected on the bend in the capillary (see Figure 9.6). While part of the light escapes or is trapped in the fused silica, another part is transmitted through the solution in the bore of the capillary by multiple total reflection on the silica/air surface. Signal-to-noise ratios were found to be improved strongly compared to a conventional optical system. Capillaries with a Z-shaped detection window are commercially available. The capillaries are pre-aligned and fixed in a capillary holder that can be obtained for several instruments. The length of the transverse section of the capillaries is 3 mm. This is somewhat on the large side for CE; generally, plate numbers do not exceed 100,000 with these Z-cells [12].

For the gain in signal-to-noise ratios with a Z-cell, values between 3 and 20 have been reported; the gain factor seems to depend partly on the detection wavelength. From Figure 9.6 it will be clear that the guiding of the light through the cell by refraction and reflection is a delicate matter. As Abbas and Shelly have shown [13], the refractive index of the solution has a large influence on the optical properties of axially illuminated capillaries. It is possible that part of the sensitivity obtained with Z-shaped optical cells is based on refractive index changes rather than on absorption of light. If this would be the case, the wavelength selectivity of UV detection would be partly lost with these devices.

9.5 Multi-wavelength detection

As in HPLC, multi-wavelength UV detection can be applied in CE for peak identification purposes or to assess peak purity. CE instruments with fast-scanning

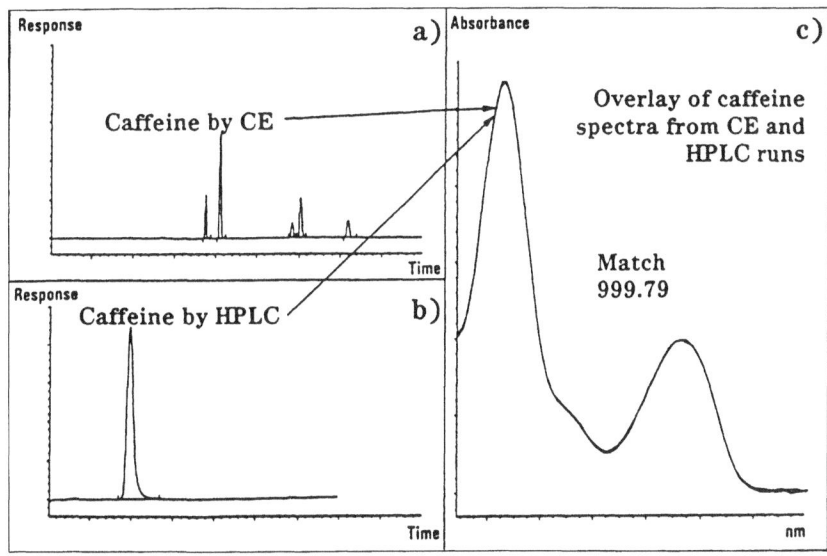

Figure 9.7

Comparison of DAD spectra of caffeine acquired by (a) CE and (b) HPLC. (c) gives the overlay of the spectra recorded at the peak maxima. Reproduced from reference [15].

detectors and diode-array detectors (DAD) have been brought on the market. The requirements for the rate of data collection are more severe in CE. With migration times of a few minutes and plate numbers of several hundreds of thousand, a total peak width of less than a second is not unusual. To obtain a reasonable number of data points or spectra for a peak, the cycle time should be less than 100 ms. With fast scanning detectors, it may be necessary to limit the scanning range. In a direct comparison of a fast-scanning with a diode-array based multiwavelength detector in CE, it was shown that the fast-scanning apparatus yielded lower detection limits when less than 8 different wavelengths were monitored; when a higher spectral resolution was required, the DAD performed better [14]. With a DAD, the cycle time is typically 12.5 ms; it has been shown experimentally that this is fast enough to allow the registration of peak shapes and spectra without bias [15].

With modern multi-wavelength detectors for HPLC, signal-to-noise ratios are similar to those obtained with a variable wavelength detector. With CE detectors this is also possible, albeit at the expense of some loss in spectral resolution. The spectral resolution of a DAD is related to the window or slit width in the axial direction of the capillary. Narrower slits, while improving optical resolution, yield lower signal-to-noise ratios. For most applications, however, the typical 2 nm resolution of HPLC detectors is not really necessary. Figure 9.7 shows that with a slightly lower resolution (7 nm) spectra obtained in CE can still perfectly match those obtained with a HPLC detector.

9.6 Thermo-optical detection

Thermo-optical absorbance (TOA) detection has been developed as an alternative for conventional UV absor-

bance detection for the measurement of very low absorbances [16]. With thermo-optical detection, the problem of conventional absorbance detection, namely that a small decrease of a relatively high background signal has to be measured, is avoided. In TOA detection two light beams are directed on the optical cell, a pump beam and a probe beam. The wavelength of the intense pump beam is selected to match the absorption wavelength of the analyte(s) of interest. When an analyte is present in the optical cell, the energy of the absorbed light is converted by radiationless transitions into heat, which causes an increase of the solution temperature. The thermo-optical effect is proportional to the absorbance by the analyte and to the power of the pump beam. To distinguish between the absorbance-induced temperature changes and overall temperature drift, a pulsed laser is preferably applied for the pump beam. Since the relaxation time for temperature changes in a CE capillary is less than a millisecond, pulse frequencies of several hundreds of Hertz can be applied.

The periodic changes of the temperature of the solution are accompanied with changes of its refractive index (RI). The RI changes can be observed with the help of the second, probing light beam. The wavelength of this beam is chosen such that it is not absorbed by the components in the solution. However, the deflection and diffraction of the probing beam at the solution-capillary interface will be disturbed in a periodic fashion by the RI modulations, which can be monitored with a small photodetector at some distance from the capillary. By coupling the modulated output of this photodetector with the frequency of the pumping beam source, with the help of a lock-in amplifier, the real thermo-optical effect can be filtered from interferences from outside.

The first application of the TOA principle in CE was presented by Bruno et al. [17]. They could measure

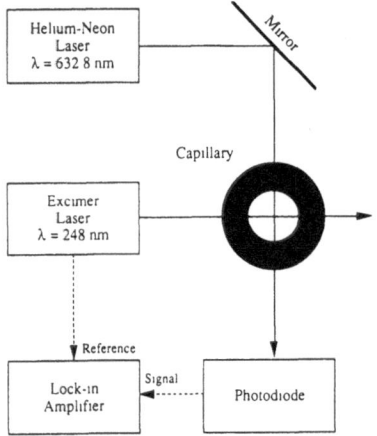

Figure 9.8
Set-up for thermo-optical absorbance detection. Reproduced from reference [18].

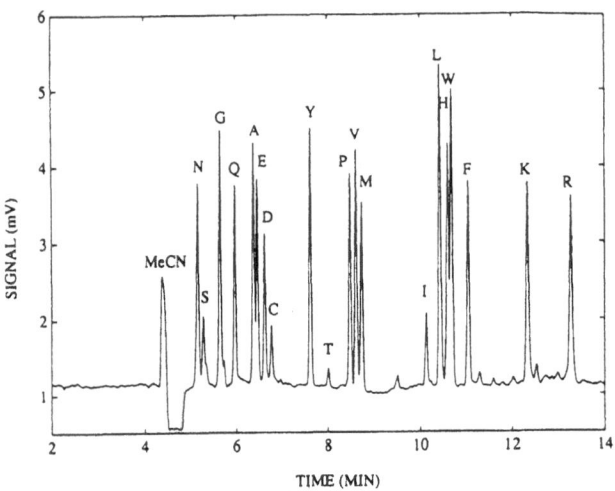

Figure 9.9
High-speed MEKC separation of PTH-amino acids with TOA detection. Reproduced from reference [18].

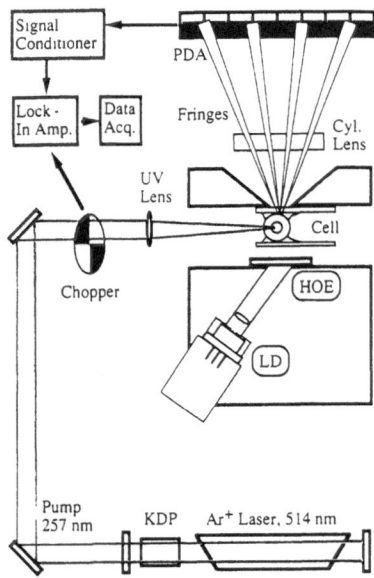

Figure 9.10
Experimental set-up for thermo-optical absorbance detection with a hologram-based RI detector. KDP = frequency doubling crystal; LD = laser diode; HOE = holographic optical element; PDA = photodiode array. Reproduced from reference [20].

dansylated amino acids, with detection limits in the 10^{-4} mol L^{-1} range, using a frequency-doubled argon ion laser as pumping source. A real break-through was realised by Waldron and Dovichi [18]. They used the 248 nm line of a pulsed KrF eximer laser for pumping, and a He-Ne laser (632.8 nm) for the probe beam. A scheme of the set-up is shown in Figure 9.8. PTH-derivatized amino acids, separated in a MEKC system, could be determined with detection limits in the order of 10^{-6} mol L^{-1} (see Figure 9.9).

Further improvements in the performance of TOA detection were realised by the research group at Ciba-Geigy in Basel [19,20]. They combined the TOA principle with a hologram-based RI detector for CE developed previously [21]. With a holographic optical element, the probe beam from a laser diode or a He-Ne laser was split in two arms, one transversing the solution inside the capillary and one the fused-silica wall. The interference pattern after recombination of the two arms was monitored with a photodiode array. A frequency-double argon ion laser was used for pumping. The set-up is shown schematically in Figure 9.10. With this detection system, proteins could be detected with a sensitivity similar to that obtained with laser-induced fluorescence detection. However, as the authors point out, the present system has a larger field of application, because not all proteins do show a substantial native fluorescence.

References

[1] S. Hjerten, Chromatogr. Rev., 9 (1967) 122.
[2] G.J.M. Bruin, G. Stegeman, A.C. van Asten, X. Xu, J.C. Kraak, and H. Poppe, J. Chromatogr. 559 (1991) 163.
[3] Y. Wahlbroehl and J.W. Jorgenson, J. Chromatogr., 315 (1984) 135.
[4] T. Wang, R.H. Hartwick, and P.B. Champlin, J. Chromatogr., 462 (1989) 147.
[5] Hewlett-Packard Co., Peak, 2 (1993) 11.
[6] R.-L. Chien and D.S. Burgi, Anal. Chem., 64 (1992) 489A.
[7] Y. Xue and E.S. Yeung, Anal. Chem., 66 (1994) 3575.
[8] N.M. Djordjevic and K. Ryan, J. Liquid Chromatogr., 19 (1996) 201.
[9] I.H. Grant and W. Steuer, J. Microcolumn Sep., 2 (1990) 74.
[10] X. Xi and E.S. Yeung, Appl. Spectrosc., 45 (1991) 1199.
[11] J.P. Chervet, N. van Soest, M. Ursum, and J.P. Salzman, J. Chromatogr., 543 (1991) 439.
[12] S.E. Moring, R.T. Reel, and R.E.J. van Soest, Anal. Chem., 65 (1993) 3454.
[13] A.A. Abbas and D.C. Shelly, J. Chromatogr. A, 691 (1995) 37.
[14] W. Beck, R. van Hoek, and H. Engelhardt, Electrophoresis, 14 (1993) 540.
[15] D.N. Heiger, P. Kaltenbach, and H.-J.P. Sievert, Electrophoresis, 15 (1994) 1234.
[16] R.C.C. Leite, R.S. Moore, and J.R. Whinnery, Appl. Phys. Lett., 5 (1964) 141.
[17] A.E. Bruno, A. Paulus and D.J. Bornhop, Appl. Spectrosc., 45 (1991) 462.
[18] K.C. Waldron and N.J. Dovichi, Anal. Chem., 64 (1992) 1396.
[19] B. Krattiger, A.E. Bruno, H.M. Widmer, and R. Dandliker, Anal. Chem., 67 (1995) 124.
[20] J.M. Saz, B. Krattiger, A.E. Bruno, J.C. Diez-Masa, and H.M. Widmer, J. Chromatogr. A, 699 (1995) 315.
[21] B. Krattiger, G.J.M. Bruin, and A.E. Bruno, Anal. Chem., 66 (1994) 1.

10 Fluorescence Detection

10.1 Optical arrangements for fluorescence detection

For HPLC it has amply been shown that the measurement of fluorescence is one of the most sensitive techniques for detection available. Limits of detection with fluorescence are often two to three orders of magnitude lower than with UV absorbance detection. Since in CE with its small dimensions it is to be expected that detection sensitivity may be problematic for many applications, especially for trace analysis, it seems obvious to develop or modify a fluorescence detector for use in CE. In fact, in the first example of modern CE shown by Jorgenson and Lukacs [1], a home-made fluorescence detector with a mercury lamp was applied. In one of the earlier commercially available instruments both a UV and a fluorescence detector were built in [2]. In home-built R&D instruments with a fluorescence detector various light sources such as deuterium, mercury, mercury-xenon, and pulsed xenon lamps have been used as the excitation source, with wavelength selection by filters or gratings. For the detection of the emitted light often a photomultiplier tube is used, although a photodiode can also be applied [2].

The lowest detection limits in CE have so far been obtained with laser-induced fluorescence detection. Still, the performance of fluorescence detection with a conventional excitation source, modified or designed for CE, is often not impressive. Detection limits, even for strongly fluorescent compounds, are generally in the order of 10^{-6} mol/L, which is only slightly better than with UV detection. The cause of this disappointing performance is twofold. First, the sensitivity, the intensity of the emitted light collected on the photosensor, may be considerably lower than in larger-scale separation methods such as HPLC. When a zone of a fluorescent compound passes the detection window, the expected signal is approximately given by:

$$S = 2.3k \cdot I_0 \cdot V_{ill} \cdot \Phi_i \cdot \varepsilon_i \cdot c_i \qquad (10.1)$$

where k is the collection efficiency of the optical set-up (the fraction of the emitted light directed onto the photosensor), I_0 the excitation light intensity and V_{ill} the illuminated volume of the solution in the optical cell. Φ_i, ε_i and c_i are the fluorescence quantum yield, the molar absorptivity and the concentration of the analyte of interest, respectively. In principle, the disadvantage of the small volume scale in CE (the low value of V_{ill} required to preserve the separation efficiency) could be counteracted by an increase of the intensity of the excitation light on the detection window. However, it is technically difficult to focus the light of a lamp sufficiently to obtain a high value of I_0, and to collect at the same time a considerable fraction of the emitted light. The advantage of using a bubble cell capillary, originally intended for UV absorbance detection, with an expanded illuminated volume, has been shown clearly for both lamp [3] and laser [4] light sources.

The other cause of the moderate performance of (lamp based) fluorescence detection in CE is the relatively high noise level encountered. The cylindrical geometry of the optical cell (the detection-window part of the capillary) is not optimal for fluorescence measurements. The light scattered from the air-silica and the silica-solution interfaces can give a high background light intensity on the photodetector, and an concomitant high noise level on the baseline signal. Of course, much of the scattered light can be intercepted with the proper slits and/or filters. Still, it appears that this is not enough. A number of optical arrangements have been developed to reduce the influence of light scattering. With some commercial fluorescence detectors (originally intended for use in HPLC) a special CE option is available. With this, the detection window of the capillary is inserted in a rectangular quartz cell, that can be filled with a liquid such as glycerine Since the refractive index of glycerine is close to that of fused silica, reflection on the outer walls of the capillary is strongly reduced. Light scattering from the walls of the rectangular quartz cell is much easier blocked from the photodetector.

Focusing of the excitation light can be realised with a ball lens system as for UV absorption detection. As was shown by Caslavska et al. [5], a commercial ball-lens cell intended for UV detection can be modified in relatively simple way to accommodate fused-silica optical fibres that collect light emitted by fluorescent analytes under a right angle from the excitation beam. A schematic representation of this cell is shown in Figure 10.1. UV absorbance and fluorescence can be monitored simultaneously with this detector. A tenfold gain in detection limits with the fluorescence mode has been shown for a number of native fluorescent compounds in body fluids.

Good results have been obtained with a collinear arrangement [6], as shown in Figure 10.2. Here, the light from a (laser) source is directed on the capillary by means of partially transparent mirror and a microscope objective. With the same objective, a part of the emission light is collected. The emission light passes the semitransparent mirror to the photodetector. By using appropriate filters the interference of reflected excitation light can be eliminated largely. In a commercially available instrument also a colinear arrangement is applied. Light from a laser source is led with optical fibres into the detection cell. The detector can be used with different laser types.

An alternative approach to diminish light scattering is the use of a sheath flow cell as promoted by the group of Dovichi [7, 8]. With a sheath flow cell, scattering of light from the walls of the capillary is completely avoided by measuring the fluorescence outside the capillary. As the end-side electrode compartment a rectangular flow-through cell with quartz windows is used. The excitation light beam is focused by means of a lens or microscope

0009-5893/00   S-59-07 $ 03.00/0

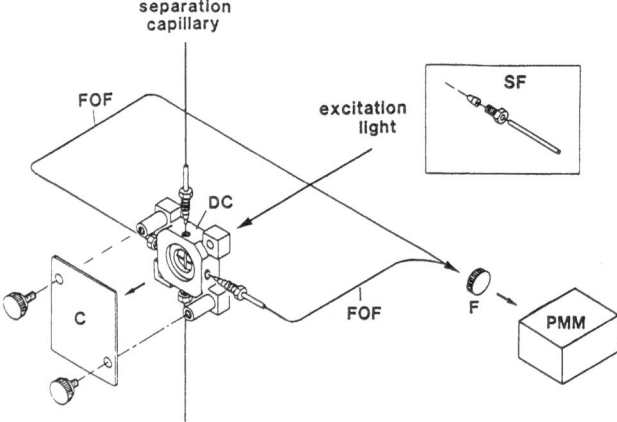

Figure 10.1
Fibre-optical cell for simultaneous absorbance and fluorescence detection. C: cover plate with photodiode for absorption measurement; DC: detector cell body; F: filter; FOF: fused-silica optical fibres; PMM: photomultiplier for fluorescence measurement; SF: sleeve and ferrule. Reproduced from ref. [5].

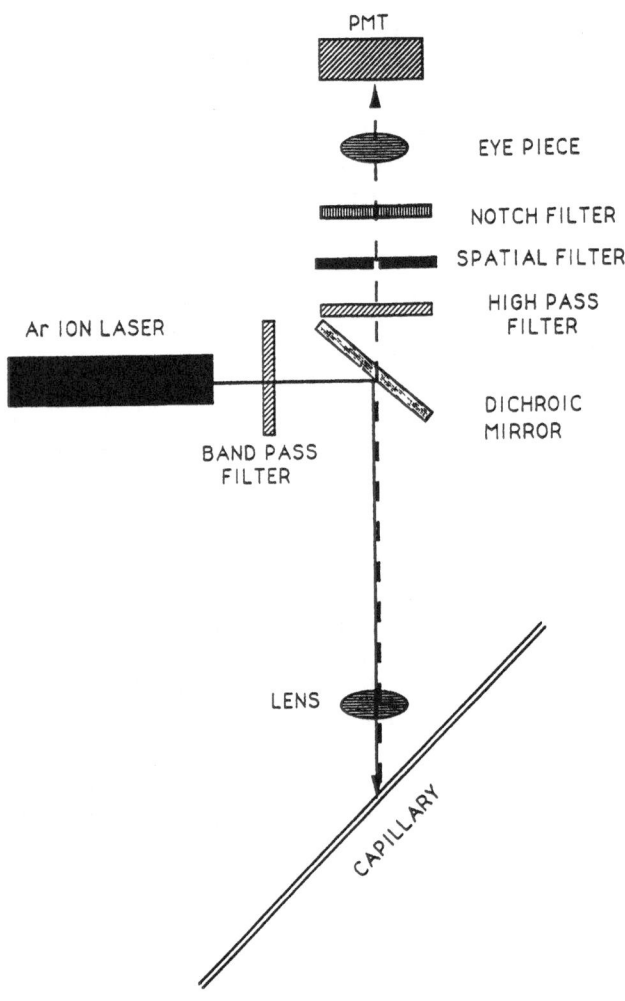

Figure 10.2
Collinear optical cell for fluorescence detection. Reproduced from ref. [6].

objective on the solution just below the end tip of the capillary (see Figure 10.3). With a second microscope objective emitted light from a limited volume of the solution is collected. In this set-up the effective volume of the 'detection cell' is not determined by the geometry of the flow cuvette, but by the optical arrangement. Without a sheath flow around the capillary end, this end-column detection mode would lead to a strong deterioration of the separation efficiency. Since the electric field outside the capillary fades away rapidly, the residence time of fluorescent analytes in the illuminated volume would be much too long to keep up with the speed of the separation. Therefore, with a pump an auxiliary liquid flow is maintained in the cuvette, to promote the flushing of the effective cell volume. The detection zone broadening can be kept negligible in this way. The sheath flow cell can be used with a laser as excitation source. With LIF, spectacular low limits of detection have been obtained. However, a conventional lamp can also be used. With a 75 W xenon arc lamp as the excitation source the detection limit for fluorescein was in the order of 10^{-9} mol L^{-1} injected [9].

In a recently developed fluorescence detector for CE interference from scattered light is prevented by separating the position of the excitation from that of the collection of the emitted light over several millimetres along the axis of the capillary. A schematic diagram of the optical set-up is shown in Figure 10.4. Upon excitation of fluorescent analyte molecules, by a beam perpendicular to the capillary axis, fluorescence light is emitted in all directions. A fraction of the this light, emitted with a direction below a critical angle to the capillary axis, will be totally reflected at the silica-air surface. Therefore, a considerable part of the fluorescence light is trapped within the capillary; the fraction of trapped emission light is approximately 15 % to either side. At some distance towards the end of the capillary (5–10 mm), a conical shaped piece of quartz or Plexiglas is placed around the capillary. This cone is brought in optical contact with the outer surface of the fused silica capillary by means of a film of glycerine. Since the refractive indices of the fused silica, the glycerine and the cone material are similar, the emission light is decoupled in the cone: rays can escape from the capillary. From the cone-air surface they are reflected to an optical fibre, that transmits the light to a photomultiplier. In Figure 10.5 the electropherogram is shown of the separation of OPA-derivatized amino acids, in a concentration of 10^{-6} mol L^{-1} each, obtained with this new detector [10]. The signal-to-noise ratios are considerably better than those obtained with a modified HPLC detector; we have found approximately two orders of magnitude lower detection limits. For OPA-derivatives, they were in the order of 10^{-8} mol L^{-1}.

A novel direction in research on LIF detection in CE is the work on emission wavelength discrimination. For this, the collected emission light is diffracted on a grating and detected with a charge-coupled device (CDD)

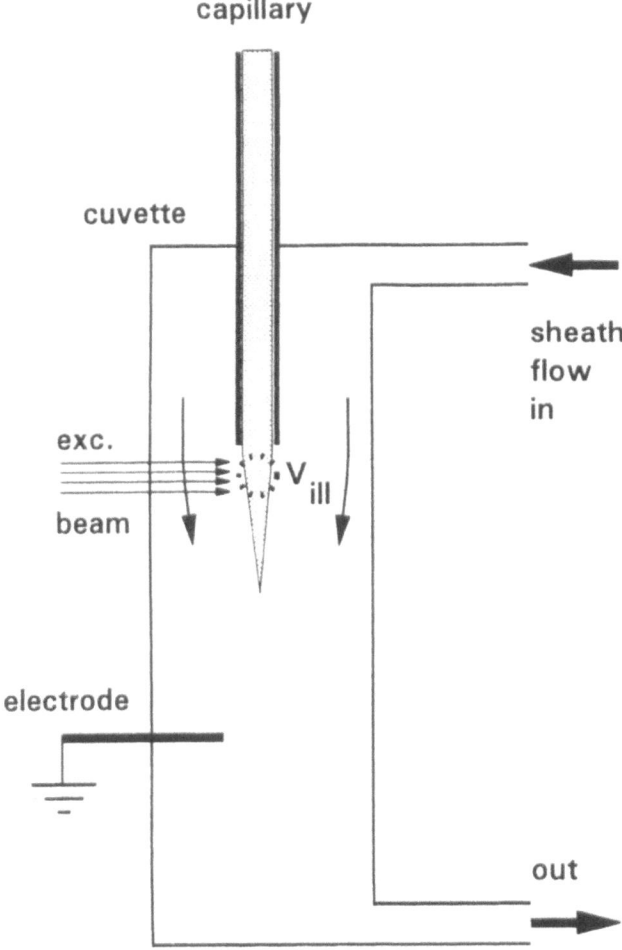

Figure 10.3
Schematic of a sheath-flow cell for fluorescence detection.

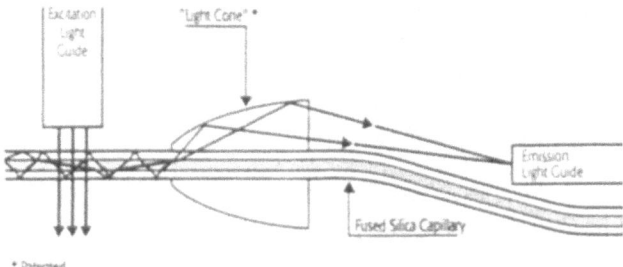

Figure 10.4
Optical principle of a fluorescence detector for capillary systems. Figure by courtesy of Flux Instruments (Pfaffenhofen, Germany).

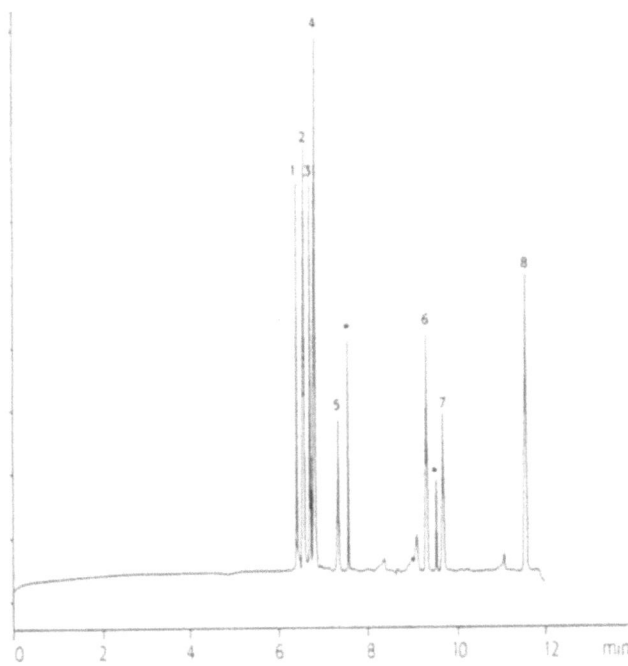

Figure 10.5
Separation of OPA derivatives of amino acids with (lamp-based) fluorescence detection. Sample concentration: 10^{-6} mol L^{-1} of each compound. For details see ref. [10].

[11, 12] or a linear diode array [13]. The main application of this technique seems to be in DNA sequencing. After the separation of the DNA sequencing fragments, the four different dye-labeled primers can be distinguished on basis of their spectral properties.

10.2 Laser-induced fluorescence detection

Laser-induced fluorescence detection (LIF) has been introduced in CE by Gassman et al. in 1985 [14] and by Burton et al. in 1986 [15]. The use of the intensive and directional light beam of a laser source for fluorescence detection in narrow capillaries had already been shown in liquid chromatography [16]. To use laser sources for fluorescence detection in CE seems therefore obvious; still, a number of patents have been obtained in the United States on this combination.

The advantages of the application of lasers are clear. The light produced by a laser is monochromatic, so that the excitation selectivity can be optimal. The parallel beam of a laser is easy to focus on a capillary, so that a high irradiance can be realized in the detection cell. The power of the laser to be used for LIF does not have to be particularly high. A line power of one or a few milliwatts is sufficient to obtain the highest sensitivities. With a higher applied power, signal-to-noise ratios are generally not further improved. This is because of different factors. First, with very high light intensities, the fluorescent compounds tend to get saturated, i.e., the majority of the molecules is already present in the excited state. An increase of the light intensity is then not giving a higher response. This saturation effect is most likely to occur with lasers that concentrate their power in short, very intense pulses. A second factor is the possible photodegradation of fluorescent analytes. In principle every fluorecent molecule may absorb and emit a photon several times during the time is is present in the illuminated detection volume. However, with some compounds the absorption of a photon may also lead to the decomposition of the molecule. When this is likely to occur, there is an upper limit to the emission intensity.

Table 10.1 Applications of laser induced fluorescence detection

Laser type	line (nm)	application[a]
Ar-ion (frequency doubled)	257	FMOC derivs.
Ar-ion (full frame)	275	proteins (native)
Ar-ion (air cooled)	457	CBQCA derivs.
	488	fluoresceine derivs.
		NBD-F derivs.
		APTS derivs.
He-Cd	325	OPA derivs.
		dansyl derivs.
		ANTS derivs.
	442	NDA derivs.
		CBQCA derivs.
He-Ne	543	rhodamine derivs.

[a]ANTS: 8-aminonaphtalene-1,3,6-trisulfonic acid
APTS: 1-aminopyrene-3,6,8-trisulfonic acid
CBQCA: 3-(4-carboxybenzoyl)-2-quinoline carboxaldehyde
FMOC: fluorenylmethoxycarbonyl
NDA: naphtalene-2,3-dicarboxaldehyde
OPA: 0-phthalaldehyde

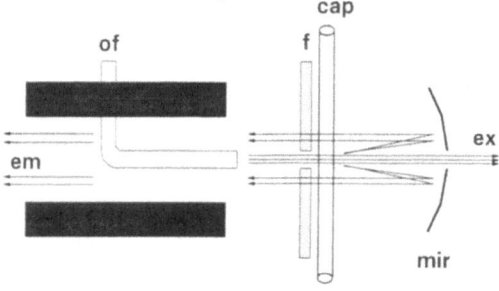

Figure 10.6
Optical arrangement for LIF detection in a commercial instrument (Beckman Instruments, Fullerton, CA). cap: capillary; ex: excitation beam; em: emmision light; f: filter/beam block; mir: ellipsoidal mirror; of: optical fibre from laser source.

For photosensitive analytes it is often better to use a less intense excitation light beam.

A third factor that limits the signal-to-noise ratios when the intensity of the excitation light is increased, is the instability of the light source. In fluorescence detection, the baseline signal is usually determined by the light scattering from the capillary walls and Rayleigh- and Raman-scattering in the solution. With low emission intensities, the noise on the baseline is mainly shot-noise, the stochastic fluctuation of the scattering processes. In this case the signal-to-noise ratios will increase with the square root of the excitation light intensity. However, with increasing light intensities the shot noise becomes relatively less important, and flicker noise may become dominant. Flicker noise is the fluctuation of the background (scattering) signal by the instability of the light source. Since the signals as well as the (noise on) the background are proportional to the excitation light intensity, a further increase of the irra-

diance does not help any more to obtain higher signal-to-noise ratios.

So far, the lowest detection limits in CE have been obtained with LIF, with the ultimate limit of one molecule coming within reach [17, 18, 19]. However, there are also some obvious drawbacks of the use of lasers, in comparison to the use of conventional light sources. First of all there are the required infrastructure, the high costs and limited lifetime of lasers that emit in the most interesting wavelength range, the (deep) UV. Still, the fact that with a particular laser only one or a few spectral lines are available is probably the most important disadvantage. Of course, this makes a LIF detector considerably less versatile than a lamp fluorescence detector. In Table 10.1 examples are given of laser lines used for LIF detection in CE.

There is a growing interest in the use of small and cheap diode lasers for LIF detection [20]. The stability of diode lasers can be excellent when the output power is feedback controlled, and their power is sufficient for application in LIF detection. Moreover, the lifetime of diode lasers is often orders of magnitude longer than that of other, more expensive types. The main problem with diode lasers is the wavelength of the emitted light, which is in the 635–900 nm range. Fluorescence with these excitation wavelengths is a very rare phenomenon. Therefore, special derivatization reagents have to be synthesised as fluorescent lables that match the wavelength of a diode laser. When such a labelling reagent is available very low detection limits can be obtained [21]. Most of the fundamental research on LIF detection in CE has been carried out with home-made instrumentation. At present (end of 1999) there are several commercial CE instrument available that offers the option to apply LIF. The excitation light from a laser is coupled into the detection cell by means of an optical fibre. This has the advantage that different lasers can be used with the same optical set-up. The construction of a cell for LIF is shown schematically in Figure 10.6.

10.3 Pre-column derivatization

Since not many compounds show strong native fluorescence, for most applications of fluorescence detection derivatization of the compounds of interest will be required. In CE, as in HPLC, derivatization can be performed before (pre-column) or after (post-column) the separation. For the choice between the two methods, similar arguments are valid as in chromatography, considering for instance the rate and optimal conditions for the reaction, the possible occurrence of multiple reaction products or the stability of the derivatives.

Pre-column derivatization procedures from HPLC have often been applied directly in CE. In other cases special labeling compounds have been developed for CE. A reason for this can be that the derivatives should be suited for a particular type of laser to be used for LIF, with an excitation wavelength compatible with a line of the laser spectrum. An example of such a derivatization reagent is naphtalene-2,3-dicarboxaldehyde (NDA), which can be used as an alternative for the well known OPA reagent. NDA shows a strong absorbance (and high fluorescence intensity) when excited with the 442 nm line of a HeCd laser [22, 23] or with the 458 nm line of an Ar ion laser [24]. Figure 10.7 shows the electropherogram of the separation of NDA derivatives of amino acids and peptides from a single cell, obtained using LIF detection.

Another reason to choose a special derivatization reagent can be that certain electrophoretic properties of the derivatives are required. As was shown in paragraph 3.2, the complete resolution of two similar compounds is generally easier to accomplish when they migrate against the electroosmotic flow. Therefore, in a BGE with the electroosmotic flow towards the negative electrode, negatively charged derivatives have some advantage. An example of such a derivatization reagent is 1-aminopyrene-3,6,8-trisulfonate (APTS), which has been used for the derivatization of oligosaccharides [25] Since the derivatization label carries three strongly acidic groups, a wide pH-range is available for the separation of the sugar derivatives.

10.4 Post-column derivatization

While precolumn derivatization procedures developed for HPLC can often be applied directly in CE, the transfer of post-column technology from LC to CE is not an easy task. First, the small volume scale of CE makes the construction of a suitable post-column reaction system technically difficult. Secondly, because of the high separation efficiency of CE, the requirements in terms of peak broadening, not only in volume units but also in time units, are much more stringent than in LC. With typical migration times of 10–15 min and plate numbers of 200,000–300,000 in CE, peak standard deviations σ are usually in the range 1–2 s. When arbitrarily a 25 % decrease of the plate number by the post-column system is accepted, the peak width contribution of the post-column system (σ_{PCR}) should be limited to approximately 0.5–1.0 s.

Different approaches have been studied in the development of a post-column system for CE. One is the free-solution approach [26]. Here, the separation capillary ends in a relatively large, grounded cuvette with the reagent. In this set-up the detection volume is not restricted physically but by appropriate focusing of the excitation light close to the separation capillary exit. To preserve the separation efficiency a sheath flow of reagent can be applied to sweep out the analyte zones from the illuminated detection volume. This approach seems technically simple but requires careful focusing of the light beam. Also, its application is restricted to very fast derivatization reactions, since longer reaction times can only be realized at the expense of a strong zone dilution. In the other approach, capillaries are used as reactor and detection cell. The reagent solution can be added to the separation medium either by pressure or by electroosmotic processes.

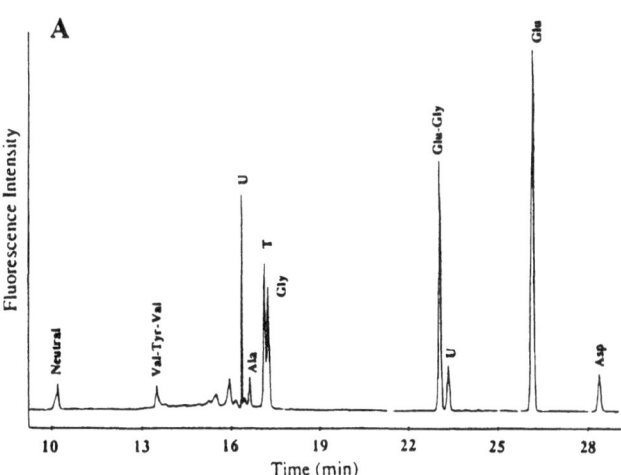

Figure 10.7
Electropherogram of the NDA-derivatized contents from a single PC12 cell, using LIF detection with a HeCd laser. Reproduced from ref. [23].

Neither the addition of the reagent by pressure nor that by electroosmotic processes are trivial. In a pressure driven system precautions must be taken to prevent laminar flow in the separation capillary, while zone broadening by laminar flow in the reaction capillary is unavoidable. In an electroosmotic driven system the flow rate of the reagent solution can not simply be regulated by the diameter ratio of the separation and the reaction capillaries, as has been suggested. In principle, in the absence of leakage currents, the electroosmotic volumetric flow rate does not change with the diameter of the capillary. In a wider piece of capillary, the advantage of a larger cross-section is exactly counteracted by a lower local field strength; since the current is the same everywhere, the field strength is inversily proportional to the cross-section. Therefore, the volumetric electroosmotic flow rate does not change at a change of the capillary diameter. To merge a reagent solution with the BGE, an extra independent voltage source would be required to control the reagent flow freely.

Apart from the technical problems to be solved, the zone broadening caused by different reaction systems will determine the choice between them. In first instance this zone broadening depends on the time required for the derivatization reaction to proceed to a certain degree of completion. Although in post-column systems a reaction yield of close to 100 % is not necessary, to gain as much as possible in sensitivity a high yield is desirable. Since in practice the reagent is always in excess, the reaction kinetics can be described by a pseudo-first order rate constant k_1:

$$Y = 1 - \exp(-k_1 \cdot t_R) \tag{10.2}$$

where Y is the yield of the reaction and t_R the reaction time. The reaction time, which should be chosen on basis of known or measured rate constants, is set by the

Table 10.2 Maximum reaction times for various capillary diameters in a pressure-driven postcolumn reaction system.

d_c (μm)	$t_{R,max}$ (s) $\sigma_{PCR} = 0.5$ s	$\sigma_{PCR} = 1.0$ s
25	27	108
50	7	27
75	3	12
100	2	7

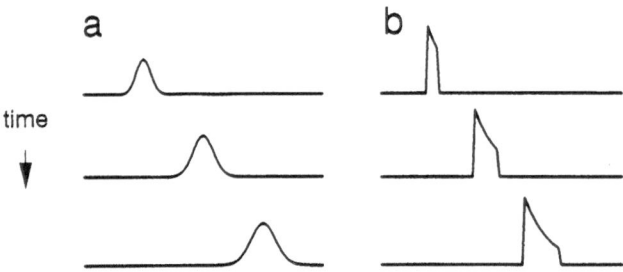

Figure 10.8
Principle of zone broadening in pressure driven (a) and voltage driven (b) post-column derivatization.

volumetric flow rate of the reaction mixture and the volume of the reactor up to the detection position.

In a pressure driven system, the zone broadening by the laminar flow in the reaction capillary can be described by the well-known Taylor-Aris equation, which can be written as:

$$\sigma^2 = \frac{d_c^2}{96D} \cdot t_R \qquad (10.3)$$

where σ^2 is the increase in the zone variance (in s^2), d_c the diameter of the reaction capillary and D the diffusion coefficient of the (derivatized) analyte ion. In Table 10.2 some examples are given of the maximally allowed reaction times with different capillary diameters. In principle, the length of the reaction capillary is not of importance. If for practical reasons a certain length has to be used, the actual reaction time and the resulting zone broadening can be decreased by increasing the reagent flow. Of course, this is accompanied by dilution of the analyte zones.

It seems obvious to use very narrow reaction capillaries to confine the zone broadening. However, with a lamp fluorescence detector with limited focusing possibilities, the illuminated volume, and with that the sensitivity, will decrease when the diameter of the capillary at the detection spot is decreased.

In an electroosmotic driven system, the osmotic flow with its flat profile will not contribute to the zone broadening. Here, however, the reaction dispersion contribution has to be considered. In general, a derivatization reaction will change the charge or size of the

analyte ion, giving a difference in the electrophoretic velocity between the non-reacted analyte and the derivatization product in the reaction capillary. In Figure 10.8 the zone broadening effect of the flow profile in pressure driven post-column reaction systems and that of the differential migration in voltage driven systems are shown schematically.

For an electroosmotically driven reagent flow, the zone broadening contribution of the reaction dispersion with relatively short reaction times ($k_1 t_R \leq 1$) is approximately:

$$\sigma^2 = \frac{1}{12} \cdot \left(\frac{\Delta \mu_{eff}}{\mu_{eff}}\right)^2 \cdot t_R^2 \qquad (10.4)$$

where t_R is the reaction time for the analyte, μ_{eff} the effective mobility of the derivatization product (including the electroosmotic contribution) and $\Delta \mu_{eff}$ the difference between the non-reacted and the reacted species. For long reaction times ($k_1 t_R > 5$) the zone broadening approaches a constant value, given by:

$$\sigma^2 = \frac{1}{k_1^2} \cdot \left(\frac{\Delta \mu_{eff}}{\mu_{eff}}\right)^2 \qquad (10.5)$$

A comparison of the performance of pressure and voltage driven post-column systems is difficult, since the first is mainly determined by the reactor diameter and the second by the chemistry of the derivatization reaction. However, as a general rule it can be stated that voltage driven systems have an advantage when a low reaction yield ($k_1 t_R < 1$) is acceptable, while a pressure driven system is preferable for the highest sensitivities.

In the group of Jorgenson a post-column reactor has been developed consisting of coaxial capillaries [27, 28]. A small O.D. separation capillary was fixed inside a $100\,\mu$m I.D. capillary which served as the reactor. A relatively high sheath flow of reagent was driven by helium pressure. In the group of Zare a cross- or tee-shaped connector was developed through which the reagent could be added with little extra zone broadening [29]. The reagent was driven by the pressure induced by a height difference. Albin et al. [30] used electroosmosis to drive the reagent – separation buffer mixture through a reaction capillary by placing the grounded electrode after the reaction capillary. The reagent was introduced through a small gap between the separation and the reaction capillary.

In our lab a pressure driven post-column derivatization system has been developed as shown schematically in Figure 10.9 [3]. In this system, the separation capillary ends in a sleeve of porous PTFE inside a reagent vessel containing the grounded end-side electrode of the high-voltage source. A second capillary, the detection capillary, is inserted in the other side of the porous sleeve. By applying a pressure on the inlet vial and the reagent vessel simultaneously, reagent solution can be merged with the electroosmotic flow without introduction of a

a

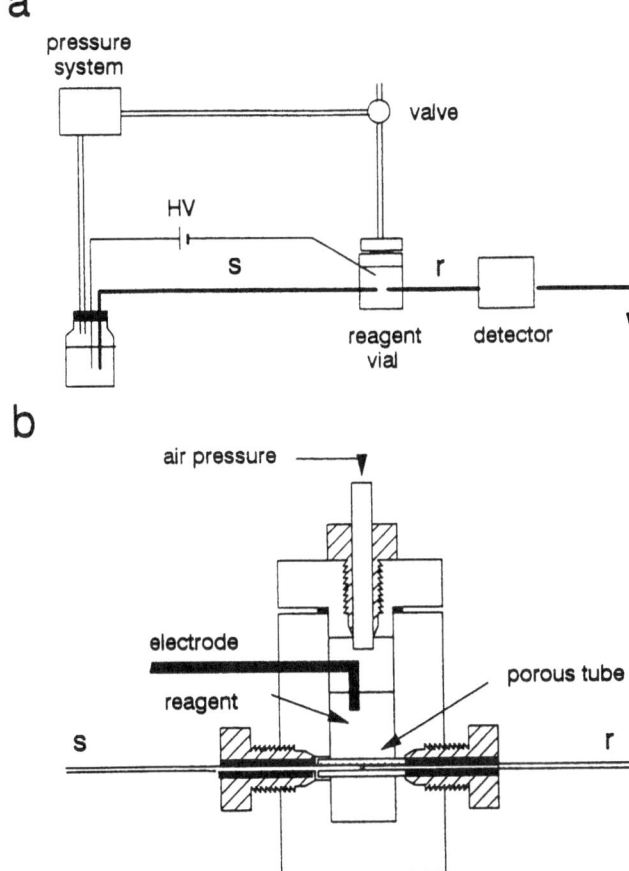

b

Figure 10.9
Schematic of a set-up for post-column derivatization in CE. Reproduced from ref. [3].

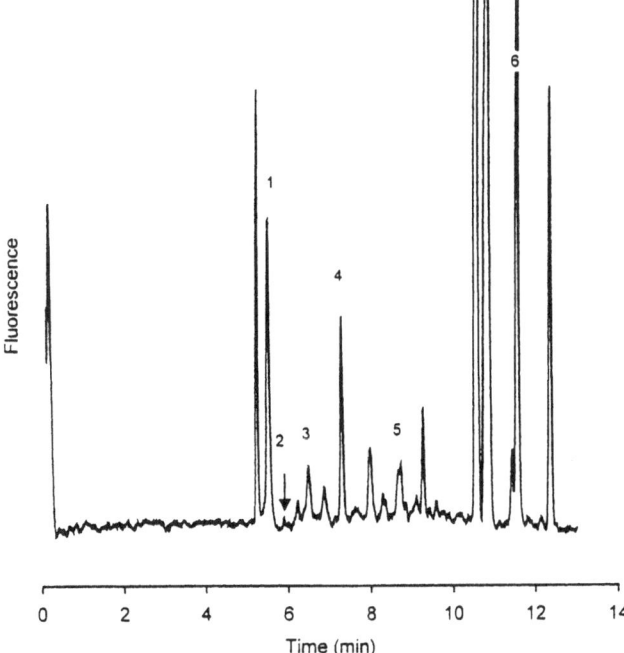

Figure 10.10
Determination of catecholic compounds in a (10 times preconcentrated) urine sample by CE with post-column derivatization and sensitised luminescence detection. 1: DA; 2: E; 3: NE; 4: DOPAG; 5: DOPA; 6: DOPAC.

laminar flow in the separation capillary. With reaction times in the order of 10 – 20 s plate numbers of over 100,000 could be obtained. The system was used for the determination of amino acids with OPA derivatization and for catecholamines and related compounds with TbCl$_3$ as derivatization reagent for sensitised luminescence detection. An electropherogram for the latter application with a urine sample is shown in Figure 10.10.

References

[1] J.W. Jorgenson and K.D. Lukacs, Anal. Chem., 53 (1981) 1298.
[2] R.G. Brownlee and S. W. Compton, Am. Lab., 20 (1988) 10.
[3] R.H. Zhu and W.Th. Kok, J. Chromatogr. A, 716 (1995) 123.
[4] R.O. Cole, D.L. Hiller, C.A. Chowojdak, and M.J. Sepaniak, J. Chromatogr.A, 736 (1996) 239.
[5] J. Caslavska, E. Gassmann, and W. Thormann, J. Chromatogr. A, 709 (1995) 147.
[6] L. Hernandez, J. Escalona, N. Joshi, and N. Guzman, J. Chromatogr., 559 (1991) 183.
[7] Y.-F. Cheng and N.J. Dovichi, Science, 242 (1988) 562.
[8] S. Wu and N.J. Dovichi, J. Chromatogr., 480 (1989) 141.
[9] E. Arriaga, D.Y. Chen, X.L. Cheng, and N.J. Dovichi, J. Chromatogr. A, 652 (1993) 347.
[10] Application Note nr. 1, Flux Instruments, Pfaffenheim, Germany.
[11] Y.F. Cheng, R.D. Piccard, and T. Vo-Dinh, Appl. Spectrosc., 44 (1990) 755.
[12] J.V. Sweedler, J.B. Shear, H.A. Fishman, R.N. Zare, and R.H. Scheller, Anal. Chem., 63 (1991) 496.
[13] S. Carson, A.S. Cohen, A. Belenkii, M.C. Ruiz-Martinez, J. Berka, and B.L. Karger, Anal. Chem., 65 (1993) 3219.
[14] E. Gassman, J.E. Kuo and R.N. Zare, Science, 230 (1985) 813.
[15] D.E. Burton, M.J. Sepaniak and M.P. Maskarinec, J. Chromatogr. Sci., 24 (1986) 347.
[16] E.J. Guthrie, J.W. Jorgenson and P.L. Dluzneski, J. Chromatogr. Sci., 22 (1984) 171.
[17] D.Y. Chen, K. Adelhelm, X.L. Cheng, and N.J. Dovichi, Analyst, 119 (1994) 349.
[18] A.T. Timperman, K. Khatib, and J.V. Sweedler, Anal. Chem., 67 (1995) 139.
[19] D.Y. Chen and N.J. Dovichi, Anal. Chem., 68 (1996) 690.
[20] A.J.G. Mank, H. Lingeman, and C. Gooijer, Trends Anal. Chem., 15 (1996) 1.
[21] A.J.G. Mank and E.S. Yeung, J. Chromatogr. A, 708 (1995) 309.
[22] B. Nickerson and J.W. Jorgenson, HRC J.High Resolut. Chrom., 11 (1988) 533.
[23] S.D. Gilman and A.G. Ewing, Anal. Chem., 67 (1995) 58.
[24] S.D. Gilman and A.G. Ewing, Anal. Methods Instr., 2 (1995) 133.
[25] A. Guttman, F.T.A. Chen, R.A. Evangelista, and N. Cooke, Anal. Biochem., 233 (1996) 234.
[26] D.J. Rose, Jr., J. Chromatogr., 540 (1991) 343.
[27] D.J. Rose, Jr. and J.W. Jorgenson, J. Chromatogr., 447 (1988) 117.
[28] B. Nickerson and J.W. Jorgenson, J. Chromatogr., 480 (1989) 157.
[29] S.L. Pentoney, Jr., X.H. Huang, D.S. Burgi and R.N. Zare, Anal. Chem., 60 (1988) 2625.
[30] M. Albin, R. Weinberger, E. Sapp, and S. Moring, Anal. Chem., 63 (1991) 417.

11 Indirect Detection

11.1 Transfer ratios

One of the attractive features of CE, and of any free zone electrophoretic separation method, is the possibility to apply indirect detection for analytes that are otherwise difficult to measure with spectroscopic techniques [1,2,3,4]. Indirect detection can also be applied in various variants of liquid chromatography; there, however, it has never become popular. In CE on the other hand, indirect detection has often been applied, if only for the simple reason that alternative detectors (refractive index, conductivity) are or were not available. Examples of analyte types often detected indirectly are inorganic [5,6] and simple organic ions [7], and sugars [8,9,10]. These compounds generally do not show any fluorescence and do not absorb UV light except in the extremely short wavelength range (<190 nm).

For indirect detection, an ionic compound with advantageous detection properties (a high UV absorbance or fluorescence intensity) is used as one of the constituents of the BGE, as the so-called monitoring or visualisation ion. It appears that the presence of any analyte ion in a zone causes a change of the concentrations of the BGE ions. The passage of an analyte zone through the detector can then be detected by the accompanying change of the monitoring ion concentration. Indirect detection is inherently universal and non-selective.

In some introduction-level texts the mechanism of indirect detection is explained as the displacement of one monitoring ion by one analyte ion of the same charge, with reference to the electroneutrality principle. This model, however, is an oversimplification. This can immediately be concluded from the fact that it is generally also possible to apply indirect detection for analyte ions with a charge opposite to that of the monitoring ion. Applications where positive and negative ions were detected in a single run, with one monitoring component in the BGE, have been shown.

For a better understanding of the principle of indirect detection and a well-founded choice of the operating conditions, the transfer ratio for a specific analyte-BGE combination should be considered. (Some authors use the term displacement ratio instead.) The transfer ratio TR can be defined as the change of the concentration of the monitoring species (A) by the presence of an analyte (i) in a zone, relative to the analyte concentration:

$$TR = -\frac{\Delta c_A}{c_i} \tag{11.1}$$

(Here, the minus sign is added for convenience, since in most practical applications a decrease in the baseline signal is registered.) Since changes of c_A are monitored, the transfer ratio is a measure of the sensitivity that will be obtained. The transfer ratio for an analyte-BGE combination can be predicted theoretically by solving

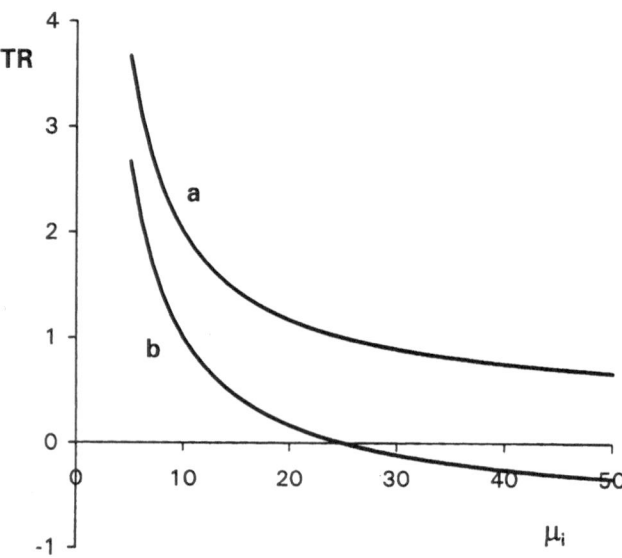

Figure 11.1

The transfer ratio TR for (single charged) analyte ions as a function of their ionic mobility μ_i in a 1:1 BGE. (a): TR of the co-ion and (b) TR of the counter ion. Assumed mobility values: $\mu_A = 25$, $\mu_B = 50$.

the appropriate equation set describing the electrophoretic transport processes. For the more simple cases, the solution can be written in terms of Kohlrausch functions, as treated in Chapter 7.

Let us first look at a simple case, with a 1:1 salt of the monitoring ion A and counter ion C as the BGE, and a strong ion i (with a charge ±1) as the analyte. The Kohlrausch equation for this case is:

$$\frac{c_i^Z}{\mu_i} + \frac{c_A^Z}{\mu_A} + \frac{c_C^Z}{\mu_C} = \frac{c_A^B}{\mu_A} + \frac{c_C^B}{\mu_C} \tag{11.2}$$

The superscripts Z and B in this equation denote the analyte zone and the BGE, respectively. Recognising that $c_A^Z - c_A^B = \Delta c_A$, and taking the electroneutrality principle into account, the transfer ratio can be found as [11]:

$$TR = \frac{\mu_A \cdot (\mu_i + \mu_C)}{\mu_i \cdot (\mu_A + \mu_C)} \tag{11.3}$$

This equation shows that the transfer ratio depends on the mobilities of all ions involved, including that of the counter ion in the BGE.

In Figure 11.1 (curve a) the transfer ratios for ions with a charge of ±1 are plotted as a function of their mobility. The X-axis in this figure can be seen as a transformation of the time axis in the expected electropherogram when a mixture of analytes would be separated, and the Y-axis as a measure of the expected peak heights. From the figure it can be seen that only for co-ions with a specific mobility the simple 1:1 displacement model is valid. Also it can be seen that "slow" analyte ions are detected with a higher sensitivity than "fast" ions. A further consequence of the Kohlrausch rules is that multiple

0009-5893/00   S-66-07 $ 03.00/0

charged analyte ions are detected with a higher sensitivity than single charged ions.

Since for most analyte ions the displacement is not on a 1:1 basis, the counter ion concentration in the zone will also be changed. Therefore, the counter ion can be used as the monitoring ion. The transfer ratio for the counter ion C (with single charged analyte ions) is also shown in Figure 11.1 (curve b). With such a scheme both negative and positive peaks may appear in the electropherogram. However, as can be seen in the figure, the absolute value of the transfer ratio for the counter ion is generally lower than that for a co-ion (an ion with the same sign as the analyte ions). Therefore, this scheme is usually not to be recommended.

As shown above, the transfer ratio for a particular analyte ion depends on the mobilities of the monitoring and counter ions in the BGE. In Figure 11.2 this dependency is shown graphically for (single charged) analyte ions with various mobilities. It can be seen that:

- monitoring ions with a high mobility give higher transfer ratios; this has also been found in practice [12];
- the use of a counter ion with a high mobility gives an advantage for "slow" analyte ions;
- with a "slow" counter ion the sensitivity for different analytes will be more uniform over the electropherogram.

As was shown in Chapter 7, Kohlrausch functions such as Equation 11.2 can also be used (within certain restrictions) to describe the electrophoretic behaviour of monoprotic weak acids and bases. In this case, the values of the ionic mobilities and the total (analytical) concentrations of the compounds in question should be substituted in the formulas. Since Equation 11.3, describing the transfer ratio, was directly obtained from a Kohlrausch equation, all conclusions derived above for strong single charged ions are equally well valid for (monoprotic) weak analytes and BGE components. This implies that the transfer ratios in a particular BGE for weakly acidic or basic analytes depend only on their ionic mobilities, and not on there pK_a values. Also it follows that transfer ratios obtained with a weak acid or base as monitoring ion, do not depend on the pH of the BGE. Therefore, one can manipulate the pH of the BGE freely to optimise the resolution of analytes without worrying about transfer ratios.

Of course, in the interpretation of Figures 11.1 and 11.2 some care should be taken when applied to weak acids and bases. The X-axes can not be translated directly into a position in the electropherogram; the migration time of a weak protolyte does depend on its pK_a and the pH of the BGE. Moreover, the transfer ratio does not reflect directly the sensitivity to be expected when the monitoring compound is a weak acid or base. It could be that the spectroscopic properties (molar absorptivity or fluorescence quantum yield) of the different ionic species of the monitoring compound differ. In such a case the sensitivity will depend on the BGE pH, albeit for all analytes in the same way.

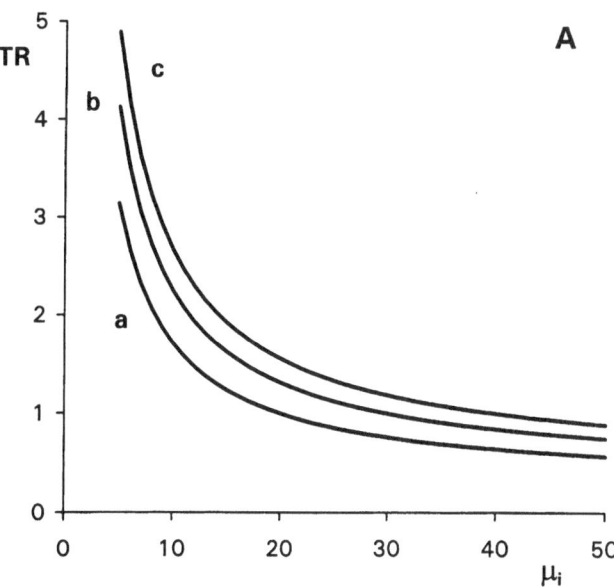

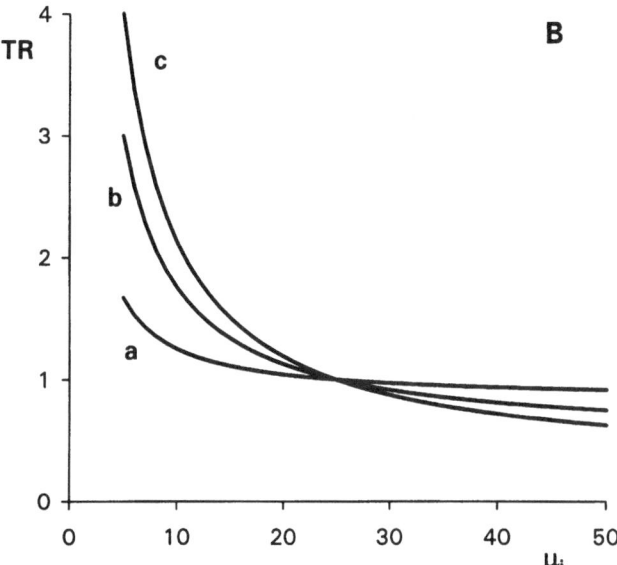

Figure 11.2
Influence of the mobilities of the monitoring and counter ions in the BGE on the transfer ratios. TR with (A) different monitoring ion mobilities μ_A of 20 (a), 30 (b) or 40 (c), and $\mu_C = 50$; and (B) different counter ion mobilities μ_C of 5 (a), 25 (b) or 75 (c), and $\mu_A = 25$.

Optimisation of the BGE composition in respect to the transfer ratios is of importance in trace analysis. When lower detection limits are not so important because the expected analyte concentrations are fairly high, it is often better to optimise the BGE in respect to the sample capacity. This may lead to quite different conclusions.

11.2 System zones

The moving boundary theory originally developed by Kohlrausch in the previous century [13] was later generalised to include multiple ion systems. It was shown that in a solution containing n different ions, (n-1) boundaries may develop during electrophoresis, one of them stagnant in the solution and (*n*-2) moving. This principle is also valid for moving zones in CE. When for instance a sample zone containing an analyte ion i (and counter ion C) is introduced in a BGE of a simple salt with ions A and C, the total number of different ions is three. There will be two zones, one of them stagnant at the original position of the sample solution, and one moving (the analyte zone). Of course, when there is an electroosmotic flow in the capillary, the "stagnant" zone will be transported through the capillary by it. Since the concentration of A in the stagnant zone will generally be different from that in the BGE, the passage of this electroosmotic peak will be visible in the electropherogram when A is a monitoring ion for indirect detection.

When the sample with analyte ion i is injected in a BGE that contains an additional ion, for instance an ion B with the same sign as the monitoring ion A, the occurrence of three zones is predicted. One of these zones is the analyte zone, one is the stagnant (electroosmotic) zone, and the third, moving zone is a so-called system zone. It is not related to the analyte ion; when a blank solution is injected, containing only the same ions as the BGE, the system peak may still be visible. The mobility of the system zone is not equal to that of any of the ions present in the BGE. For the example above, its mobility is somewhere between that of ion A and B, depending on the concentrations of A and B in the BGE:

$$\mu_{sys} = \frac{\mu_B(\mu_A + \mu_C) \cdot c_A + \mu_A(\mu_B + \mu_C) \cdot c_B}{(\mu_A + \mu_C) \cdot c_A + (\mu_B + \mu_C) \cdot c_B} \quad (11.4)$$

The occurrence of system zones is not limited to indirect detection systems. However, with direct detection of the analytes the system zones are often not visible in the electropherogram, because they consist of a migrating disturbance of the (non-detectable) BGE constituents.

The intensity of a system zone is related to the extent in which the sample matrix differs from the BGE composition. Such differences are often large in comparison to the analyte concentrations in the sample. System peaks are therefore often high and broad due to overloading phenomena. When an analyte zone migrates with a mobility close to that of a system zone, its shape may be distorted and its presence difficult to detect because of the interference of the system zone.

Since the number of system zones expected in an electropherogram is related to the total number of different ions in the BGE, it can be stated as a general advise that with indirect detection the BGE composition should be kept as simple as possible. What should also be kept in mind is that an extra system peak can be expected when the pH of the BGE is low (<4) or high (>10). Under these conditions the contribution of the H_3O^+ or OH^- ions to the charge transport can not be neglected, so that they count as an extra ion type in the BGE.

The presence of additional ions apart from the monitoring ion in the BGE has also an influence on the transfer

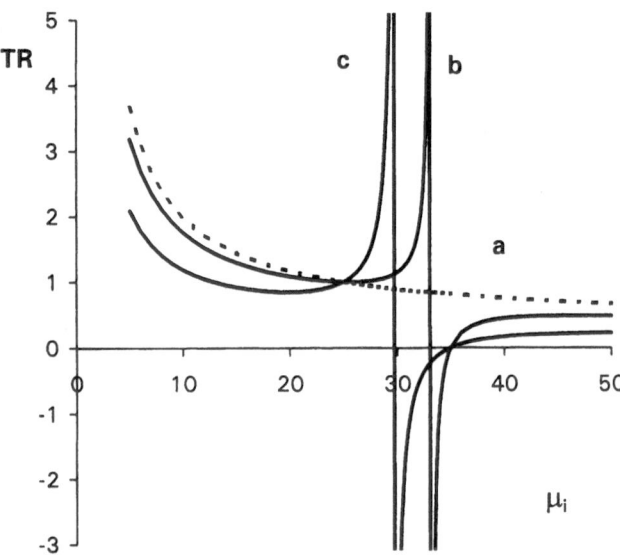

Figure 11.3
Transfer ratios for the monitoring ion A in a BGE with an additional co-ion B. Concentration of the additional co-ion: (a) $c_B = 0$; (b) $c_B = 0.2\ c_A$; (c) $c_B = c_A$. Assumed mobility values: $\mu_A = 25$, $\mu_B = 35$ and $\mu_C = 50$.

ratios for analyte ions. In Figure 11.3 the displacement of monitoring ion A is shown as a function of the analyte mobility, when an additional ion B (with the same sign as A) is present in the BGE. The TR values have been calculated for various ratios of the concentrations of A and B in the BGE. The dotted line is the transfer ratio for a BGE without additional ions. It can be seen that the presence of B may have a drastic influence on the transfer ratio. While for most analyte mobilities the predicted sensitivity will be decreased, in a part of the electropherogram very high positive or negative peaks are expected. In principle it is possible to tune the BGE composition, by the addition of extra salts with the appropriate ionic mobilities, in such a way that for a particular analyte a very high sensitivity will be obtained. Unfortunately, this is always in a mobility region close to the position of a system zone. Interference of the system zone is then always to be expected. Moreover, the mobility range for which a high sensitivity is expected is very narrow, while its position depends on the exact composition of the BGE (see Figure 11.3). A slight change of the BGE composition might change the sensitivity for a particular analyte strongly, sometimes even with a reversal of the direction of the peak. It will be clear that in such a case the robustness of the method is not very satisfactory.

It has been suggested that high transfer ratios can be obtained in different parts of the electropherogram by using a BGE with a mixture of monitoring ions. In Figure 11.4 the calculated effect of this approach is shown. In the figure the total displacement of three monitoring ions with different mobilities is given as a function of the analyte mobility. Indeed, a high sensitivity is predicted in various parts of the electropherogram. However, the drawbacks as mentioned

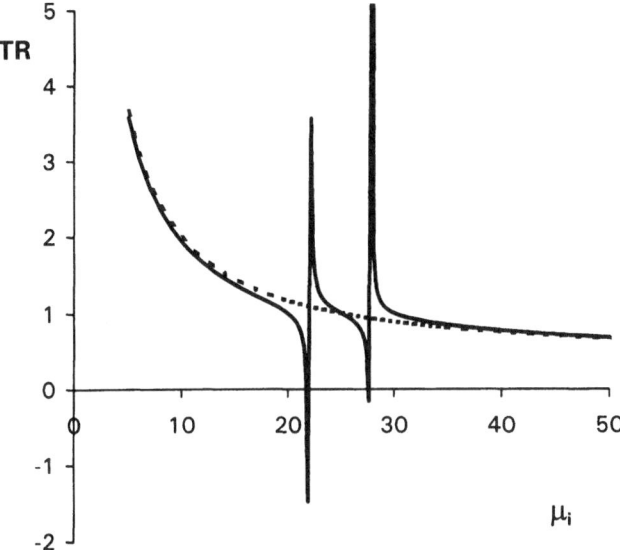

Figure 11.4
Transfer ratios in a BGE with a mixture of monitoring ions A1, A2 and A3 in equal concentrations. Assumed mobility values: $\mu_{A1} = 20$, $\mu_{A2} = 25$ and $\mu_{A3} = 30$; $\mu_C = 50$. The dotted line gives the transfer ratios with only A2 as monitoring ion.

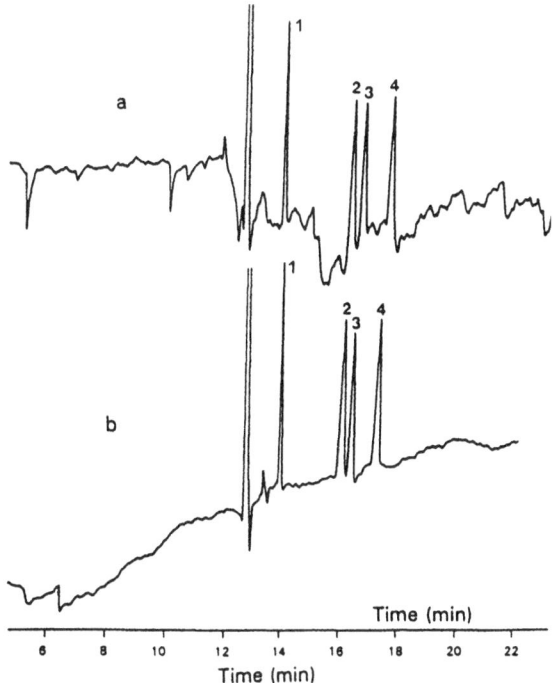

Figure 11.5
Influence of the capillary inner diameter on the baseline stability in the indirect detection of sugars. BGE: 12 mmol L^{-1} sorbate, 63 mmol L^{-1} NaOH. Peaks: 1 mmol L^{-1} of sucrose (1), maltose (2), glucose (3) and fructose (4). Reproduced from reference [10].

above are still valid; the regions of increased sensitivity are very narrow, their positions depend on the exact composition of the BGE and they are always close to a system peak. Therefore, the practical utility of this approach seems to be limited.

11.3 Baseline instabilities in indirect detection

In practice it is often found that the quantitative reliability of indirect detection for low analyte concentrations is not so much limited by the detector noise but rather by instabilities of the baseline. Baseline drift and shifts and the appearance of unexpected, often broad peaks in the electropherogram may interfere with the recognition and integration of the analyte peaks. A remarkable feature of these baseline instabilities is that they are often repeatable and seem to be unrelated to the injected sample. In fact, it is usually not even necessary to inject any sample to obtain the same baseline pattern after switching on the high voltage source.

The degree of baseline instability is related to the Joule heat development during the separation. When it is not possible to use a low conductivity BGE, for reasons of sample capacity or because a high or low pH is required for the separation, the detectability can be improved by using a smaller diameter capillary. An example of this is shown in Figure 11.5. In the indirect detection of sugars, separated at a high pH, a much more stable baseline was obtained with a 25 μm ID capillary than with a 50 μm capillary. Decreasing the separation voltage may also help, of course at the expense of a longer analysis time.

Another important experimental parameter for the baseline stability is the operating temperature. Especially when the thermostating conditions vary over the

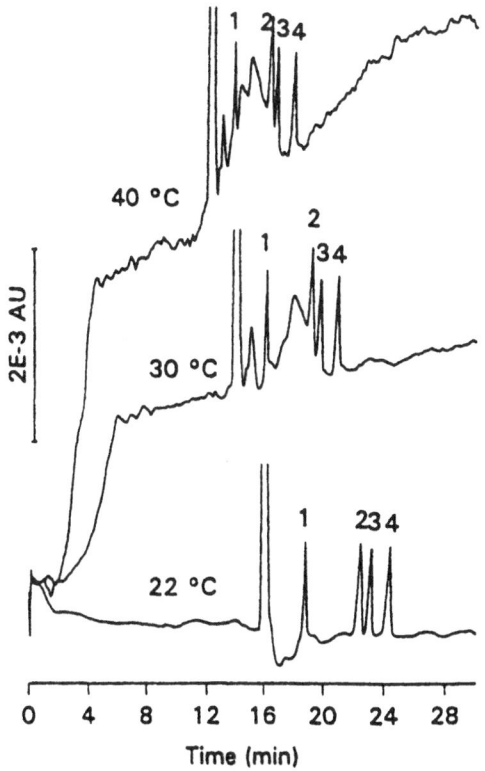

Figure 11.6
Influence of the operating temperature on the baseline stability in the indirect detection of sugars. Conditions as in Figure 11.5. Capillary inner diameter 25 μm. Reproduced from reference [10].

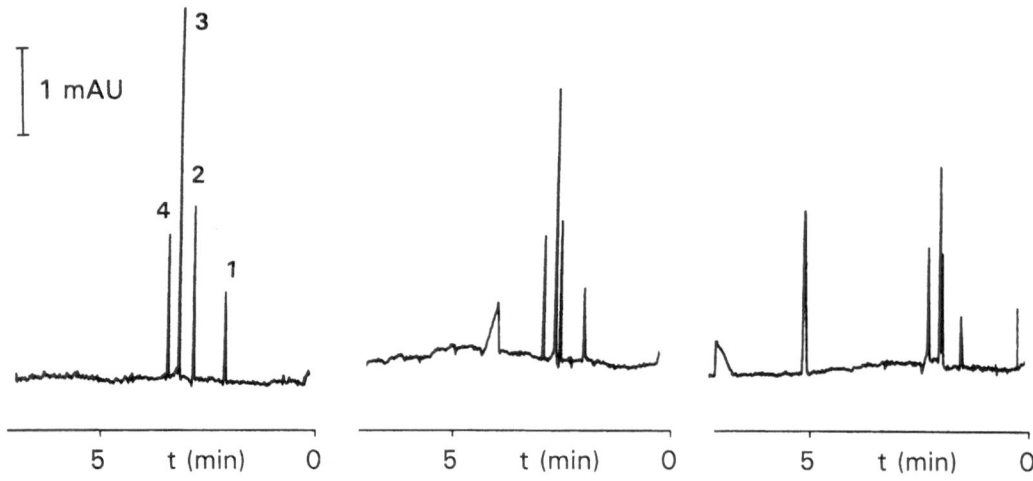

Figure 11.7

Influence of the BGE concentration on the signal-to-noise ratios in indirect UV detection. Monitoring ion: pyridine, detected at 255 nm. Monitoring ion concentration of 5 (a), 2.5 (b) and 0.5 (c) mmol L^{-1}. Peaks: 0.2 mmol L^{-1} of K$^+$ (1), Na$^+$ (2), Mg^{2+} (3) and Li$^+$ (4).

length of the capillary, the baseline instabilities increase with the difference between the set-point and ambient temperatures (see Figure 11.6). These and other observations led to the conclusion [14] that the baseline instabilities are caused by temperature differences over the length of the capillary, either already existing before the separation or induced at the start by unequal heat dissipation conditions. Baseline shifts and disturbances could be tracked down as stemming from "thermal nodes" in the instrumental set-up, migrating with the velocity of system peaks through the capillary. The disturbances of the monitoring ion concentration can come from both ends of the capillary to the detection window. To eliminate all thermal nodes from an instrumental system seems impossible. However, the number of disturbances coming from one thermal node is equal to the number of system peaks, as determined by the BGE composition. The advise to keep the BGE as simple as possible is therefore also of value in respect to the baseline stability.

11.4 Indirect UV detection

Application of indirect UV detection is very simple. For the BGE, a salt of a small, UV absorbing ion can be chosen. For the detection of anions, monitoring ions such as chromate, benzoate or sorbate can be used; for cations, imidazole and pyridine have been recommended. As has been discussed in the previous paragraphs, the addition of other ionic components to the BGE, to improve the selectivity of the separation, should be kept to a minimum.

A number of studies have been devoted to the optimisation of the signal-to-noise ratios in indirect UV detection [3,15,16]. It is generally recognised that the noise observed is composed of two contributions. One source is the detector noise. Its magnitude is virtually independent of the type and concentration of the monitoring ion, and only slightly dependent on the detection wavelength. The detector noise is mostly electronic, but with the low light intensities transmitted through the detection window in CE, shot noise may also be of importance. The other contribution was found to be proportional to the absorption by the BGE [16], i.e., to the molar absorptivity (ε_A) and the concentration of the monitoring ion. The source of this noise contribution is a real fluctuation of the monitoring ion concentration. The total noise can then be written as:

$$N_{tot}^2 = N_{det}^2 + (k \cdot \varepsilon_A \cdot c_A)^2 \tag{11.5}$$

where k is a constant indicating the magnitude of the proportional noise under specific experimental conditions. Wang and Hartwick [16] found that they could decrease the proportional noise by suppressing the electroosmotic flow. They attributed the proportional noise to adsorption and desorption phenomena. Xu et al. [14] on the other hand, attribute the proportional noise to migrating disturbances of the BGE composition, induced by small thermal "nodes" on the capillary, similar to the process causing large baseline disturbances. This view is not necessarily in contradiction with the observations of Wang and Hartwick. When in a simple salt BGE the only system zone is the stationary zone, the suppression of the electroosmotic flow will prevent that disturbances pass the detector.

For the performance of a particular indirect detection system, of course the signal-to-noise ratios obtained are decisive. In indirect UV detection the signal for a particular analyte concentration c_i can be written as:

$$S = TR \cdot \varepsilon_A \cdot L_{eff} \cdot c_i \tag{11.6}$$

where ε_A is the molar absorptivity of the monitoring ion and L_{eff} the effective light path length. To evaluate the signal-to-noise ratio, we can now distinguish two extreme situations. One case is when the detector noise is dominant and the other when the proportional noise is dominant. For the first case, the signal-to-noise ratio is:

$$S/N = TR \cdot \frac{\varepsilon_A \cdot L_{eff}}{N_{det}} \qquad (11.7)$$

Under these conditions the concentration of the monitoring ion is not of importance for the sensitivity or the signal-to-noise ratio. An example of this is shown in Figure 11.7. The separation of some inorganic cations is shown with pyridine as monitoring ion, and tartaric acid as the buffering component of the BGE. Dilution of the BGE does not give a significant improvement of the detection. With the lower BGE concentrations broadening of the analyte zones by overloading becomes significant, and peak heights are decreased.

Selecting a monitoring ion with a higher absorptivity will give better signal-to-noise ratios. However, this is valid only up to a certain limit; above this, the proportional noise will start to become dominant, and further improvements will not be obtained. It is then even possible that the signal-to-noise ratios decrease again when a monitoring ion with a higher absorptivity is chosen. As has been shown in Chapter 9, the linearity of UV detection in CE is often limited, depending on the quality of the focusing of the light on the capillary. The sensitivity of detection will be decreased at higher absorbance values. When for indirect detection a strongly absorbing BGE is used, the background absorption is easily in this non-linear region, and peak heights will be lower than expected.

With a dominant proportional noise the signal-to-noise ratio is:

$$S/N = TR \cdot \frac{L_{eff}}{k \cdot c_A} \qquad (11.8)$$

In this case the absorptivity of the monitoring ion is not of special importance. Better signal-to-noise ratio are obtained with lower monitoring ion concentrations. However, a disadvantage of diluting the BGE is that overloading of sample zones is more likely to occur (see Figure 11.7). Moreover, below some monitoring ion concentration the detector noise will become dominant again.

The influence of the monitoring ion absorptivity (ε_A) and concentration (c_A) on the signal-to-noise ratio is shown graphically in Figure 11.8. It is always best to choose a monitoring ion with a high absorptivity. When one finds oneself in the situation that the sensitivity is decreased by the non-linear behaviour of the detector (on the right side in Figure 11.8), it is always possible to decrease the background absorption by choosing a different wavelength for detection. There is no law stating that the monitoring ion should be detected at its absorption maximum. Optimisation of the monitoring ion concentration, on the other hand, implies always to find

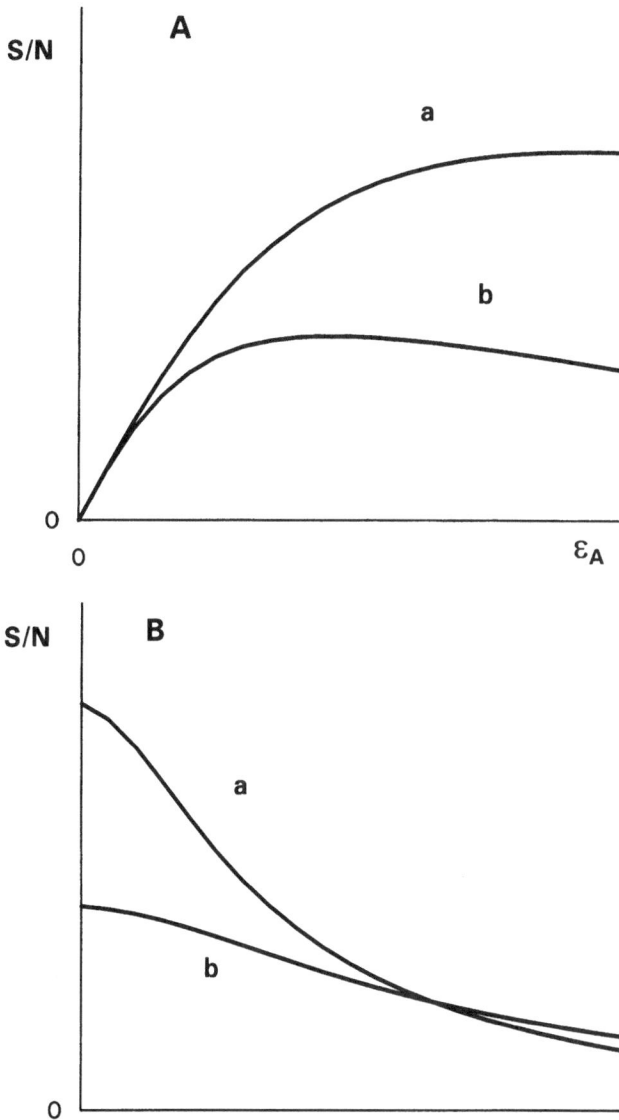

Figure 11.8
Influence of the molar absorptivity ε_A and concentration c_A of the monitoring ion on the signal-to-noise ratios in indirect UV detection. (A) Influence of ε_A with low (a) and high (b) monitoring ion concentration. (B) Influence of c_A with low (a) or high (b) absorptivity of the monitoring ion.

the best compromise between a low limit of detection (low c_A) and a high sample capacity (high c_A).

11.5 Indirect fluorescence detection

Indirect fluorescence detection has been explored mainly by the group of Yeung [2,17,18]. The principle behind it is exactly the same as with UV detection: a monitoring ion present in the BGE is displaced by analyte ions, and the passage of a zone through the detector is noticed as a dip in the baseline signal. As with UV detection, the method is inherently non-selective. Fluorescent monitoring ions such as quinine for cations and salicylate for anions have been applied.

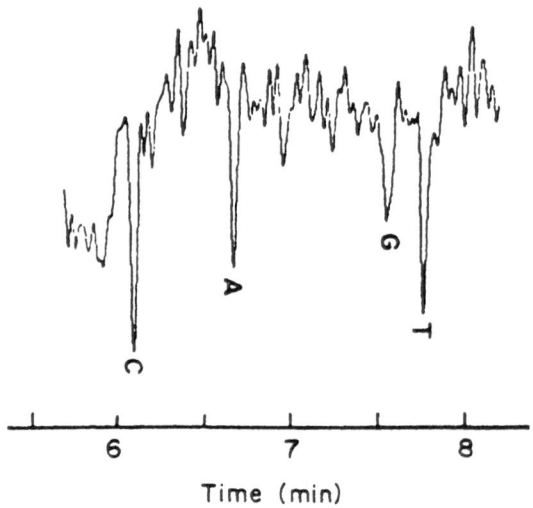

Figure 11.9

Indirect fluorescence detection of nucleotide monophosphates. Monitoring ion: salicylate, 0.25 mmol L^{-1}. Injected concentrations: $2\mu10^{-6}$ mol L^{-1}. Reproduced from reference [17].

Compared to indirect UV detection, fluorescence detection can give much lower detection limits. The cause of this difference is in the different noise characteristics of the two measurement methods. With UV detection, the intensity of the light on the photodiode is relatively high. The absorbance by the solution inside the capillary causes only a small decrease of the intensity of the transmitted light, since the absorbance has to be kept low to preserve the linear behaviour of the detector (see paragraph 9.2). Therefore, the measured light intensity is only slightly depending on the concentration or composition of the BGE, and the noise on the background signal is then also virtually independent of the operating conditions. In other words: the non-proportional noise contribution is easily dominant. This means for indirect UV detection that a decrease of the monitoring ion concentration below a certain value does not give an improvement of the signal-to-noise ratios anymore. With a fluorescence detector, the background light intensity (without fluorescent compounds in the solution) is in principle zero. Of course, this ideal condition may not be met in practice because of scattering of the excitation beam or stray light. However, by a proper design of the instrumental set-up such interferences can be kept small.

The performance of indirect fluorescence detection is not automatically improved by changing from a lamp to a laser as excitation source. Of course, with an appropriate laser much higher excitation and emission intensities can be obtained. For the signal-to-noise ratios, however, the stability of the excitation source is also of importance. The quality of the source can be expressed by its dynamic reserve (*DR*), the signal-to-noise ratio of the intensity of the light emitted by it [4]. When the fluorescence intensity of the monitoring ion is high enough, and stray and scattered light are sufficiently blocked from the detector, the level of the (proportional)

noise on the baseline is mainly determined by the dynamic reserve of the source. With a good lamp the *DR* can be as high as 10^4, while with some types of laser the *DR* may be lower than 10^2. In the earlier work of Yeung et al. on indirect fluorescence detection it was shown that a dramatic reduction of detection limits could be realised by using a (commercially available) power stabiliser for the laser source.

When indirect fluorescence detection is applied, the background signal and the noise on it are approximately proportional to the concentration of the monitoring ion (c_A), even at low values of this concentration. The limit of detection (with *S/N*=2) for a component i can be estimated as:

$$LOD = \frac{2c_A}{TR_i \cdot DR} \qquad (11.9)$$

where TR_i is the transfer ratio for the analyte. Decreasing the monitoring ion concentration may still improve the signal-to-noise ratios in indirect fluorescence detection when this would not be beneficial anymore in indirect UV detection. In practical applications one sees typical monitoring ion concentrations of 5 – 10 mmol · L^{-1} with UV and 1 mmol · L^{-1} with fluorescence detection. The detection limits are correspondingly lower [19].

Low detection limits (in the order of 10^{-7} mol/L) can be obtained with a stabilised laser source and a very low monitoring ion concentration. An example is shown in Figure 11.9. However, with such low BGE concentrations the separation system is easily overloaded by salt in the sample matrix, so that only low-conductivity samples can be analysed.

References

[1] S. Hjerten, K. Elenbring, F. Kilar, J.L. Liao, A.J.C. Chen, C.J. Siebert, and M.D. Zhu, J. Chromatogr., 403 (1987) 47.

[2] W.G. Kuhr and E.S. Yeung, Anal. Chem., 60 (1988) 1832.

[3] F. Foret, S. Fanali, L. Ossicini, and P. Bocek, J. Chromatogr., 470 (1989) 299.

[4] E.S. Yeung and W.G. Kuhr, Anal. Chem., 63 (1991) 275A.

[5] W.R. Jones and P. Jandik, J. Chromatogr., 546 (1991) 445.

[6] M. Chen and R.M. Cassidy, J. Chromatogr., 640 (1993) 425.

[7] W.R. Jones and P. Jandik, J. Chromatogr., 608 (1992) 385.

[8] A.E. Vorndran, P.J. Oefner, H. Scherz, and G.K. Bonn, Chromatographia, 33 (1992) 163.

[9] T.W. Garner and E.S. Yeung, J. Chromatogr., 515 (1990) 639.

[10] X. Xu, W.Th. Kok, and H. Poppe, J. Chromatogr., 716 (1995) 231.

[11] M.W.F. Nielen, J. Chromatogr., 588 (1991) 321.

[12] S.M. Cousins, P.R. Haddad, and W. Buchberger, J. Chromatogr., 671 (1994) 397.

[13] F. Kohlrausch, Ann. Phys. Chem., 62 (1897).209.

[14] X. Xu, W.Th. Kok and H. Poppe, in preparation.

[15] G.J.M. Bruin, G. Stegeman, A.C. van Asten, X. Xu, J.C. Kraak and H. Poppe, J. Chromatogr., 559 (1991) 163.

[16] T. Wang and R.A. Hartwick, J. Chromatogr., 607 (1992) 119.

[17] W.G. Kuhr and E.S. Yeung, Anal. Chem., 60 (1988) 2642.

[18] E.S. Yeung and W.G. Kuhr, Anal. Chem., 63 (1991) 275A.

[19] P.E. Andersson, W.D. Pfeffer, and L.G. Blomberg, J. Chromatogr. A, 699 (1995) 323.

12 Electrochemical Detection

12.1 Principles of electrochemical detection techniques

In this chapter two detection modes will be discussed that are based on the measurement of an electric current: amperometric and conductivity detection. Both modes are invasive techniques, i.e., they both require a direct contact between the separation medium and the sensing element(s), the electrode(s). In other respects there are cardinal differences between the two. Amperometric detection relies on the chemical characteristics of specific analytes, while with conductivity detection a bulk property of the separation medium is monitored. The first technique is therefore selective, the second inherently non-selective.

With amperometric detection, specific analytes are oxidised or reduced on the surface of the sensing (working) electrode. Electric charge is passed from the particles in the solution to the conducting electrode. The rate of this charge transfer process, measurable as the electric current through the working electrode, is dependent on two factors. First, the electroactive compound has to be transported from the bulk of the solution to the electrode surface, by convection and diffusion. Since the rate of this mass transport is proportional to the concentration of the substance involved, the current can be taken as a measure for the concentration of an electroactive compound in the solution. Secondly, the rate of the charge transfer itself on the electrode surface depends on the electrode potential, or better, the electric potential difference between the working electrode and the solution. When the electrode potential is high enough (or negative enough in case of a reductive process), the charge transfer is virtually instantaneous and mass transport is the rate limiting process.

Selectivity in amperometric detection is governed by the electrode potential required for the analytes of interest. With easily oxidised (or reduced) compounds, the electrode potential can be kept moderate to give the full sensitivity. Possible interference's with a higher oxidation or reduction potential will then give (much) lower signals. When a relatively high detection potential is required for a certain analyte, the selectivity of the method will be inferior [1].

The sensitivity of an electrochemical flow-through detector cell can be quantified by its coulometric yield, the fraction of the analyte actually oxidised or reduced while passing the detector. The coulometric yield is mainly a function of the cell geometry, the size of the working electrode and the flow rate of the solution. Of course, for the detection limits the noise level is equally of importance. Baseline noise generally increases in proportion with the electrode surface area.

As has been shown previously in liquid chromatography, a small detection volume is not necessarily a disadvantage for an electrochemical detector. With amperometric detection, miniaturisation of the cell volume may lead to an improved mass transfer towards the electrode surface, and an increased sensitivity. With a decreased electrode surface area improved signal-to-noise ratios can be obtained [2,3]. An amperometric detector has already been developed for open-tubular LC, which has a volume scale even smaller than CE [4].

Conductivity detection in CE relies on the non-idealities of zone electrophoresis. As has been shown in Chapter 7, the presence of an analyte ion in a zone disturbs the local conductivity of the solution. When for instance an analyte ion A migrates through a BGE with co-ion B and counter ion C, the change of the conductivity of the solution ($\Delta\kappa$) can be found, with the help of the Kohlrausch equation, as:

$$\frac{\Delta\kappa}{\kappa^{\beta}} = \frac{\mu_A - \mu_B}{\mu_A} \cdot \frac{c_A^{\zeta}}{c_B^{\beta}} \tag{12.1}$$

where κ^{β} is the conductivity of the BGE, c_A^{ζ} the concentration of the analyte in the zone and c_B^{β} the BGE salt concentration. Depending on the mobilities of the ions involved, positive and negative peaks can be expected in the electropherograms.

The changes of the conductivity, which have to be measured on a background, will be relatively large when the BGE concentration is low. On the other hand, the highest sensitivities will be obtained for analyte ions having a mobility strongly different from that of the background co-ion. For such ions overloading will readily occur when a low BGE concentration is used. Therefore, with conductivity detection one has always to look for the best compromise between sensitivity and resolution.

To measure the conductivity of a solution, a voltage is applied between two (platinum) electrodes in contact with the solution and the current between them is monitored. With conductivity detection one is interested in the bulk of the solution; the processes occurring at the electrodes are merely adventitious. To minimise the interference of the electrode processes on the signal, usually a high-frequency alternating voltage is applied. For application in CE this has the additional advantage that it makes it easier to discriminate the (AC) signal from the DC electrophoretic current.

One of the major problems for the combination of CE with an electrochemical detection technique (CE-ECD), is the interference of the high electric field used for the separation with the detection, where very small currents have to be measured. In principle, there are three positions possible for the sensing electrode(s) in CE-ECD (see Figure 12.1). The sensing electrodes can be placed somewhere along the length of the capillary, in a so-called on-line mode. This is of course the normal mode with spectrometric detection; however, in that case there is no direct contact between the sensor and the solution. On-column amperometric detection is certainly not possible, since the solution at a point somewhere along the capillary would have a potential several kV from ground. For amperometric detection a potential

0009-5893/00   S-73-07 $ 03.00/0   © 2000 Friedr. Vieweg & Sohn Verlagsgesellschaft mbH

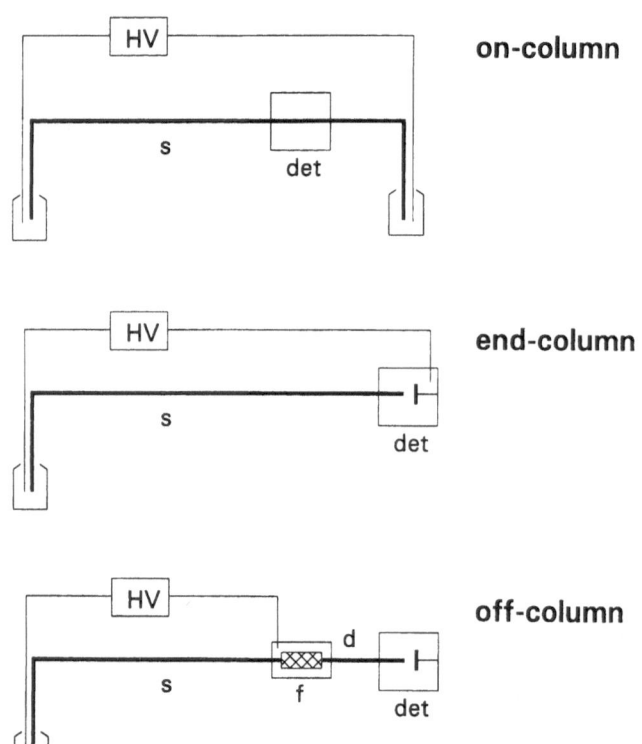

on-column

end-column

off-column

Figure 12.1
Modes for CE-ECD. s: separation capillary; d: detection capillary; f: field decoupler; det: detector; HV: voltage source.

difference of 50 mV is already significant. On-column conductivity detection has been demonstrated, with extremely small sensing electrodes and a very good protection of the AC measuring circuitry from the (DC) separation voltage [5].

A second possibility is to place the electrode(s) directly in the buffer vial at the end of the capillary, together with the grounding electrode of the separation circuitry, for so-called end-column detection [6,7]. To avoid extensive zone broadening and dilution of the separated zones before detection takes place, the electrode(s) in end-column detection should be placed very close to the exit of the separation capillary, or even brought in rectally.

The third possibility is off-column detection [8]. In this mode the electrophoretic voltage is decoupled before detection takes place. After decoupling, the solution with the separated analyte zones is transported towards the detector through a second piece of capillary. The performance and limitations of end-column and off-column electrochemical detection will be discussed in the next paragraphs.

12.2 End-column detection

The success of end-column amperometric detection strongly depends on the diameter of the capillary. Figure 12.2 schematically shows the detector end of the separation capillary in a buffer vial with the grounded

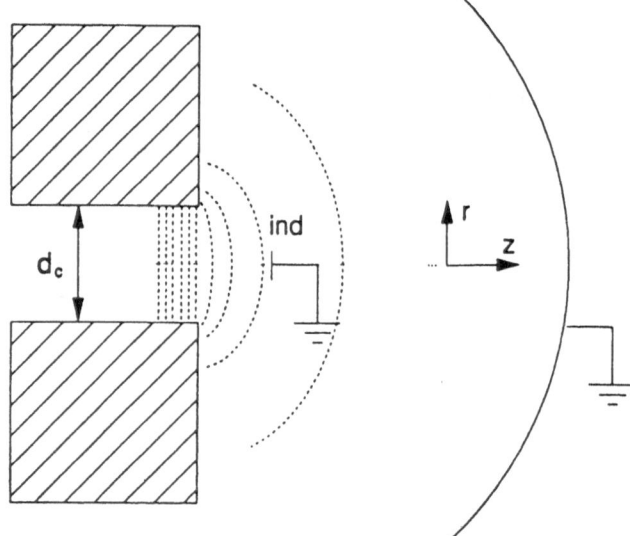

Figure 12.2
The solution potential near the capillary end in end-column detection. Dotted lines are equipotential lines.

electrode at some infinite" distance from it. A possible position of a small sensing electrode for amperometric detection is indicated in the figure. The dotted lines in Figure 12.2 are calculated equipotential lines. The lines show that the solution close to the capillary end is at a certain potential level from ground due to the ohmic drop in the buffer vial. The total voltage drop V_{tot} between the capillary end and the grounded electrode is determined by the field strength in the capillary (E_c) and the inner diameter of the capillary (d_c):

$$V_{tot} = \frac{1}{8} \pi \cdot d_c \cdot E_c \qquad (12.2)$$

In Figure 12.3 the solution potential is shown as a function of the distance of the capillary end, for different capillary diameters. The solution potential drops rapidly with the distance from the capillary end. However, this distance should be kept small, not only to have a reasonable coulometric yield or sensitivity, but also to keep the response adequately fast. Therefore, a bias of the working electrode potential can never be completely avoided in end-column detection.

A number of problems result from the potential gradients existing in the solution near the capillary end in end-column detection:

- With most potentiostats, the working electrode potential is kept virtually at ground, while the solution potential is controlled by the set-value of the detector. In the solution near the capillary end ground loops may exist, so that the real working electrode potential (or better, the potential difference between working electrode and adjacent solution) becomes dependent on the separation voltage. Moreover, the true detection potential becomes dependent on the positioning of the working electrode, and is therefore not easy to reproduce [9].

- At the onset of the separation, when the high voltage is switched on, the true detection potential is suddenly changed. Generally, this results in a strong baseline excursion, so that it becomes difficult to measure with a high sensitivity.
- The dependency of the real detection potential on the separation voltage causes any noise on the high voltage to be translated into detector noise.

It will be clear from the above that it is important in end-column detection to keep the potential gradients in the solution near the capillary end small. Apart from using low separation fields, this can be realised by using narrow capillaries. With end-column amperometric detection, the capillary diameters are generally between 5 and 25 μm instead of the usual 50 or 75 μm. The injection volumes should be decreased accordingly. Therefore, even when the mass detection limits reported can be very low, in concentration units they are often only in the order of 10^{-7} to 10^{-6} mol L^{-1}[10].

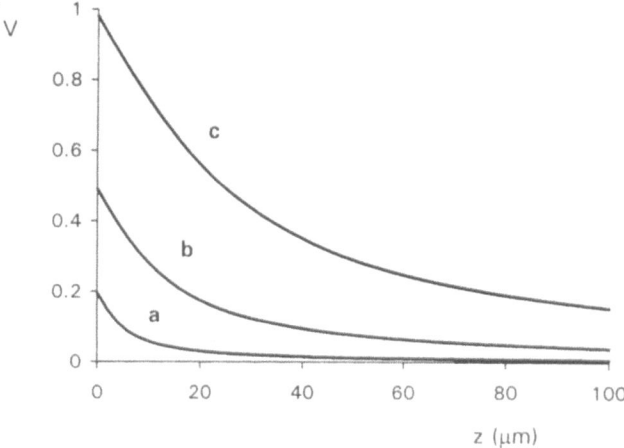

Figure 12.3

The solution potential in end-column detection as a function of the distance from the capillary end (z) and the capillary diameter. Capillary I.D.: (a) 10; (b) 25; (c) 50 μm. Electric field: 500 V/cm in the capillary.

12.3 Decoupling of the electric field

Another strategy followed to decrease the interference of the separation voltage with the detection is to decouple the electric field at some distance along the capillary before the detector. The separation voltage is then applied only between the injection end of the capillary and the point of decoupling. The propagation of the solution with the separated zones from this point on to the detector, relies on the pumping action of the electro-osmotic flow in the first part of the capillary.

Decoupling of the electric field can not be realised by simply inserting a miniature electrode through the wall of the capillary, to serve as the ground for the electrophoretic voltage. At the surface of such an electrode, hydrogen or oxygen gas would develop. Associated with the passage of the electrophoretic current, decomposition of the water in the BGE will occur at the electrodes. The evolution of gas bubbles inside the capillary would readily interrupt the current and stop the separation process. It is also not feasible to decouple the field through a hole in the capillary, or a T-junction, to an electrode in an external solution. When the electric current is branched off in this way, the EOF will also be directed along this path, and the analyte zones will not reach the detector.

Field decoupling can only be realised with the help of a conductive joint that allows the passage of the electric current but withholds the EOF. In the first work on CE-ECD [8], a porous glass tube was placed and sealed over a crack in the fused silica capillary. The complete joint was immersed in a buffer reservoir with the grounded electrophoresis electrode. Because of the small pore size (5 nm) of the porous glass material, bulk liquid flow through it is strongly suppressed, so that (most of) the EOF is directed to the detector. Figure 12.4 shows the construction and application of this decoupler. Other materials that can be used as a conductive joint over a

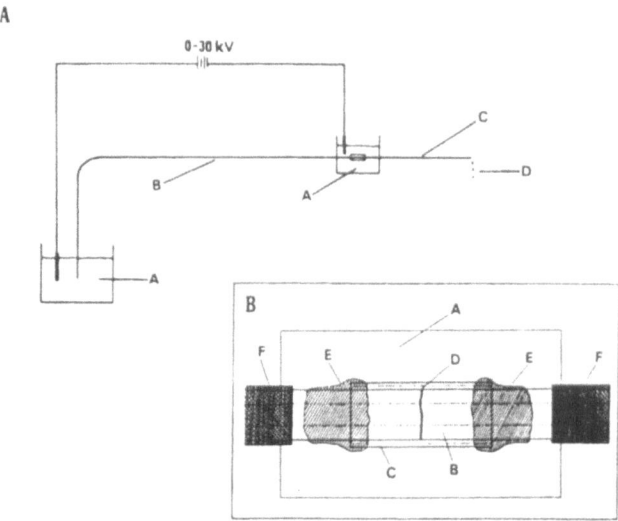

Figure 12.4

Field decoupling in CE. (A) Schematic of the system: A, buffer reservoirs; B, separation capillary; C, detection capillary; D, eluent; (B): detailed schematic of the porous joint: A, microscope slide; B, fused silica capillary; C, porous glass capillary; D, joint; E, epoxy; F, polymer coating. Reproduced from ref. [8].

fracture in the capillary are Nafion tubing [11] and porous cellulose acetate film [12].

In our laboratory we have developed another type of decoupler [13,14]. We use a union of palladium metal to couple the separation capillary and a coupling capillary leading to the detector (Figure 12.5). The palladium union is grounded and serves as the negative electrode for the separation voltage. The hydrogen produced at the palladium surface during the passage of the electrophoretic current is readily absorbed into the metal, so that bubble formation inside the capillary is avoided. The dead volume of this decoupler was shown to be only a few nanoliters.

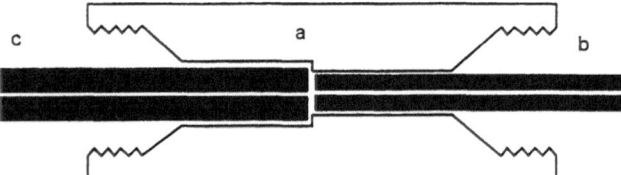

Figure 12.5
A palladium metal union as field decoupler. Figure not on scale. (a) Pd union; (b) separation capillary; (c) detector capillary. Reproduced from ref. [14].

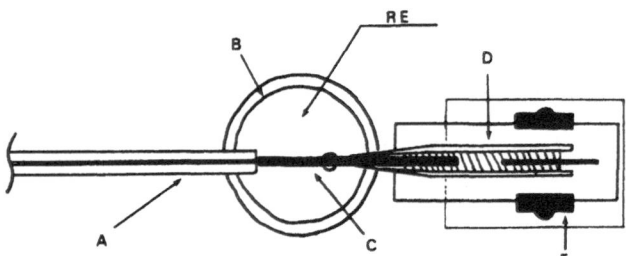

Figure 12.6
Schematic of end-column detection with a carbon fibre electrode. (A) capillary; (B) buffer reservoir; (C) carbon fibre; (D) electrode assembly; (E) micromanipulator; (RE): reference electrode. Reproduced from ref. [6].

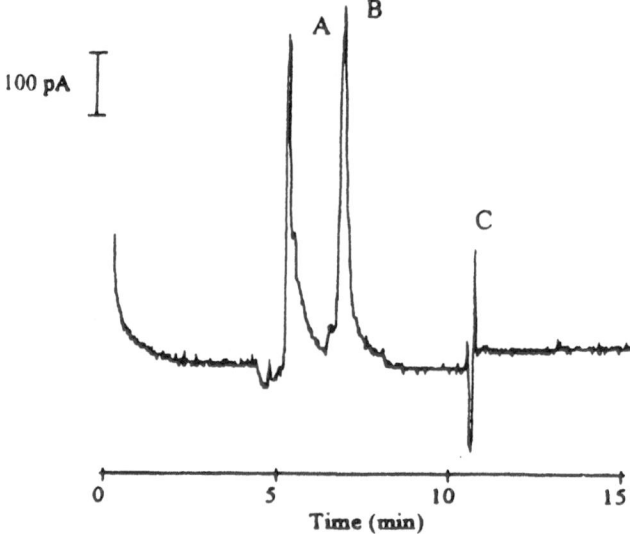

Figure 12.7
Electrophoretic separation of the components of a single dopamine neuron after injection and lysing in the capillary tip for 1 min with buffer. Peaks A and B are attributed to dopamine from different types of vesicles in the cell. Reproduced from ref. [16].

The laminar flow in the coupling capillary, where there is no electric field present to drive an EOF, will cause some extra zone broadening in off-column detection. To minimise the loss of separation efficiency a coupling capillary as short as possible should be used [8]. Moreover, the backpressure of the coupling capillary may disturb the EOF in the separation capillary itself, thereby giving rise to another increase of the zone widths. However, this distortion of the flat EOF profile can easily be eliminated by applying an adequate compensating pressure on the front end of the separation capillary during the separation [15]. The compensating pressure should be such that it is just enough to propel the (undisturbed) EOF of the separation capillary through the coupling capillary.

12.4 Amperometric detectors

So far, amperometric detectors designed for use in CE are not yet commercially available. Researchers have used home-made cells and electrodes, which were not always very user-friendly. In most early applications of amperometric detection in CE, a single carbon fibre is used as the working electrode. The fibre is mounted in a holder, and positioned in or close to the exit of the separation capillary with the help of a micropositioner. An example of such a detection set-up is shown in Figure 12.6. Because of the low detector currents obtained, it is often enough to use a two-electrode configuration, with the counter electrode serving as a quasi-reference electrode. Peak heights are in the pA range; a low-noise

amplifier and proper electromagnetic shielding of the set-up are therefore necessary. Noise levels appear to depend on the applied separation voltage, with end-column [9] as well as off-column detection [13].

With a fibre electrode the separation efficiency can be kept high, even when a narrow capillary is used for the separation. This makes CE-ECD especially suited for applications in which the available sample amount is very small, such as in single-cell analysis . An example of this type of application is shown in Figure 12.7 [16]. Here, a dopamine cell of a pond snail was sucked into a 25 μm capillary by electrokinetic injection. After an incubation period for cell lysis, separation was carried out. Scanning electrochemical detection was performed with a 5 μm diameter carbon fibre. With a short incubation time two peaks were observed in the electropherogram; both were attributed to dopamine, present in two different types of vesicles in the cell.

For routine applications carbon fibre electrodes seem less suitable. Their handling and installation is cumbersome and the set-up is not very robust. Moreover, the electrochemical characteristics of the material are not ideal so that signal-to-noise ratios are relatively low. Several research groups have investigated the use of larger scale electrodes in CE [17,18,19]. It appeared that with such electrodes, and the related larger effective detection volume, the separation efficiency is compromised to some extent. Still, plate numbers in the order of 100,000 appeared to be possible. Recent investigations in our laboratory showed that it is even possible to use a commercially available detector, intended for use in micro HPLC [20]. This detector is equipped with a 1-mm diameter glassy carbon electrode. Although the volume of the cell seemed too large for CE (25 – 50 nl),

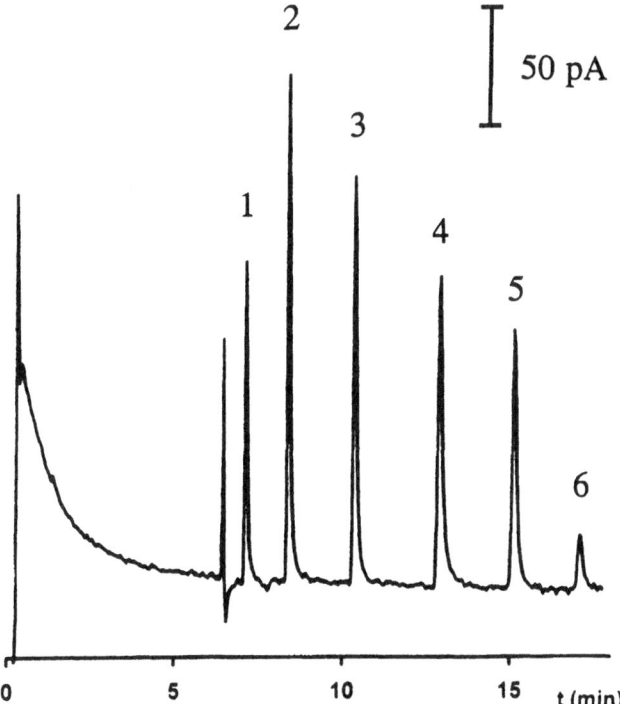

Figure 12.8
Separation of chlorophenols by CE-ECD using a commercially available detector cell. Sample: standard solution containing 5×10^{-7} mol L^{-1} of each of the compounds.

Table 12.1 CE-ECD with special active electrode materials.

Active substance	Electrode bulk material	Application	Ref.
Cu	metal	carbohydrates	28,29
		amino acids	30,31
Ni	metal	carbohydrates	32
		polyhydroxy antibiotics	33
Hg	gold	thiols	34
CuO	carbon	carbohydrates	35
CoPC[a]	carbon	thiols	36,14

[a]: cobalt phthalocyanine.

the response time was found to be in the order of $1 - 2$ s. This could be explained as the effect of the depletion of the electroactive analytes in the cell; within 1 to 2 s the conversion of the analytes at the electrode surface was complete. In Figure 12.8 an electropherogram is shown obtained with this detector. Plate numbers up to 100,000 and detection limits in the order of 10^{-8} mol L^{-1} could be obtained.

Apart from constant-potential detection other electrochemical detection modes can also be applied in CE, such as dual-electrode detection [21,22], scanning techniques [16], and pulsed amperometric detection (PAD) [23,24]. With these techniques it becomes increasingly difficult to preserve the high plate numbers typical for CE. With dual electrode detection the cell volume becomes easily too large. With PAD, and also with potential scanning detection, there is a conflict between the ideal cycle time for detection and the sampling rate required in CE [25]. For PAD, the working electrode potential follows a multistep waveform, allowing the subsequent execution of electrode cleaning, activation and measurement in a short cycle time (typically $0.5 - 1$ s). PAD is increasingly used in CE for the detection non-UV absorbing compounds such as carbohydrates [23,24] and (glyco)peptides [26,27]. Platinum or gold micro-electrodes are used, because these materials catalyse the oxidation of such compounds.

The selectivity of (constant-potential) electrochemical detection can be influenced by the choice of the electrode material. Carbon can be regarded as a general-purpose material. Certain metals or chemically modified materials can enhance the oxidation or reduction of specific classes of analytes that react slowly on carbon. For instance, copper wire electrodes can be used for the detection of compounds that form complexes with copper ions, such as amino acids and carbohydrates. Figure 12.9 shows the electropherogram for the separation of sugars, obtained with a copper electrode. In Table 12.1 a number of applications of CE-ECD with special electrode materials is listed.

12.5 Conductivity detectors

In the work on electrophoresis in narrow tubes preceding modern CE a conductivity detector was used as a standard device [37]. However, with the emergence of CE in glass or fused silica capillaries it was found that on-column spectrometric (UV absorbance or fluorescence) detection was almost ideal, and conductivity detection was neglected for a period. Still, there are some strong points for the use of conductivity detection in CE. First, the method is in principle universal and therefore especially suited for compounds that are otherwise difficult to detect, such as inorganic ions. Secondly, the small volume scale required for the detector is no impediment to obtain low detection limits.

In the first attempts to revive conductivity detection in CE, an on-column scheme was applied [38,5]. In the work by Huang et al. [5] platinum wires that served as the sensing electrodes were inserted through laser-drilled holes in the wall of the capillary. Because these electrodes were subject to the high separation voltage, they had to be aligned carefully and an isolation transformer had to be used for the measurement of the solution conductance.

In later experimental work end-column detection was applied [6,7]. One of the sensing electrodes was a platinum wire positioned right in front of the capillary exit; for the other electrode in the measurement circuitry the grounded electrode of the high voltage supply was used. With this set-up, detection limits were in the order of 10^{-5} mol L^{-1} for inorganic ions.

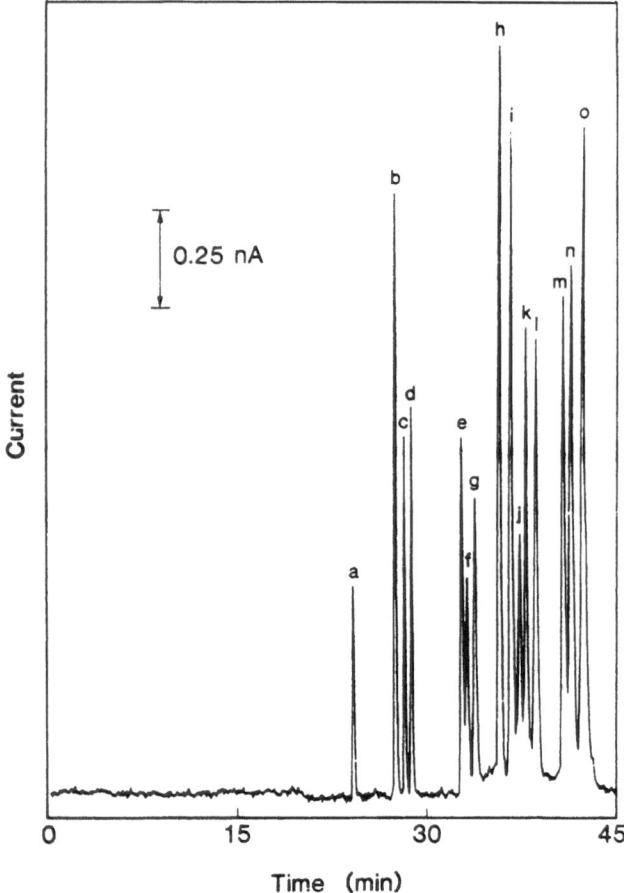

Figure 12.9

Separation of oligosaccharides by CE-ECD using a copper wire electrode. Sample: standard solution containing approximately 10^{-4} mol L^{-1} of each compound. Reproduced from ref. [28].

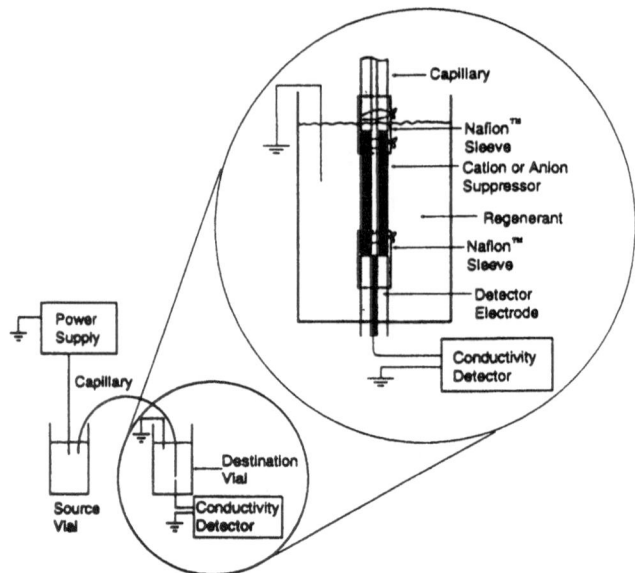

Figure 12.10

Set-up for suppressed conductivity detection in CE. Reproduced from ref. [40].

A conductivity detector with the end-column principle has recently been made commercially available [39]. With this detector the sensing electrode and the detection end of the separation capillary are permanently encapsulated in two coupling connectors, similar to those used for the coupling of optical fibres. The volume between the capillary tip and the sensor is flushed with a make-up solution towards the grounding electrode. With this so-called "open architecture" construction electrodes and capillaries are easily interchangeable, while the detection volume remains precisely defined. The flushing of the volume over the sensing electrode ensures a high response rate, even in the absence of a substantial EOF. Detection limits found were lower than with indirect UV detection.

One of the inherent problems with conductivity detection is the fact that a bulk property of the BGE is measured. Fluctuations of the BGE composition or the temperature are therefore directly translated into baseline instabilities. A high background conductivity of the BGE will be associated with a high noise level and negatively affect detection limits. With a low conductivity BGE on the other hand, the system is easily overloaded. This is the more so since conductivity de-

tection is based on the non-ideal behaviour of the BGE conductivity in CZE (see paragraph 12.1). Similar problems in ion chromatography have been diminished by the application of a suppressor system. By means of a suppressor column or membrane after the separation column, the counter ions of the mobile phase buffer were replaced by H^+ or OH^- ions. Thereby, the background conductivity and subsequently the detector noise could be strongly decreased. A prerequisite for suppressed conductivity detection is that the BGE buffering ion is weaker (a weaker acid in case of anion anaysis and a weaker base in case of cation analysis) than the analytes. Weakly acidic or basic analytes will also be neutralised in the suppression system and will therefore not be detected.

It has been shown that suppressed conductivity detection can also be applied in CE [40,41]. Figure 12.10 gives schematically the set-up that has been used. The capillary end is connected, by means of Nafion sleeves, to a 80 μm I.D. suppressor tube that goes to the sensing electrode. The suppressor tube is made of "radiation grafted" Teflon. This material can be made permeable for cations (in anion analysis) or anions (in cation analysis). A regenerant solution (sulphuric acid solution or borate buffer, respectively) surrounds the semipermeable tube; the grounding electrode is placed in this solution. For the signal-to-noise ratios the application of a suppression system has a number of advantages:

• In the suppressor tube the BGE counter ions (e.g., Na^+ or Cl^-) are replaced by H^+ or OH^- ions, which have a very high mobility. Therefore, when a (strong) analyte ion migrates from the capillary, the increase of the conductivity by the presence of this analyte ion and its counter ion is higher than without suppression.

Electrochemical Detection

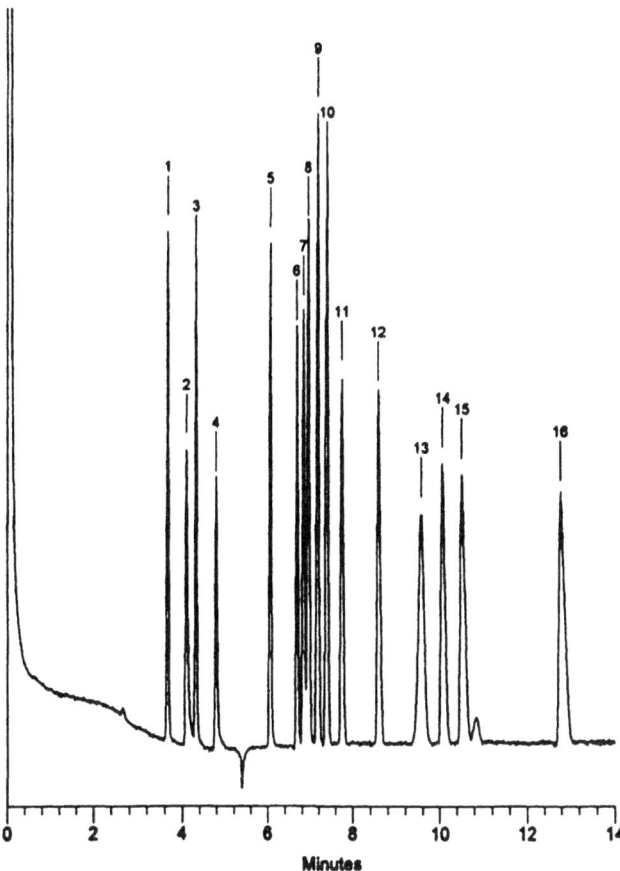

Figure 12.11
Separation of a mixture of carboxylic acids by CE with suppressed conductivity detection. Sample: standard mixture containing 10^{-5} mol L^{-1} of each compound. Reproduced from ref. [40].

- The background conductivity and baseline noise are decreased because of the neutralisation of the BGE buffer by the H^+ or OH^- ions from the regenerant solution.

- Since the suppressor tube can conduct the electrophoretic current, the separation field is largely decoupled before detection takes place. Interference from the noise on the separation voltage on the detector signal may be expected to be smaller than in normal end-column detection.

In Figure 12.11 an example is shown of suppressed conductivity detection. Detection limits for simple inorganic and organic ions are in the order of $0.2 - 1 \times 10^{-7}$ mol L^{-1} or < 10 ppb. Despite the fact that this work has been carried out in the laboratory of an instrument supplier, a suppressed conductivity detector has not been brought on the market.

References

[1] W.Th. Kok, in R.W. Frei and K. Zech (Eds.), Selective sample handling and detection in HPLC. Part A, Elsevier, Amsterdam, 1988, Chapter 6.
[2] D.M. Morgan and S.G. Weber, Anal. Chem., 56 (1984) 2560.
[3] J.W. Bixler and A.M. Bond, Anal. Chem., 58 (1986) 2859.
[4] L.A. Knecht, E.J. Guthrie, and J.W. Jorgenson, Anal. Chem., 56 (1984) 479.
[5] X. Huang, K.T. Pang, M.J. Gordon, and R.N. Zare, Anal. Chem., 59 (1987) 2747.
[6] X. Huang, R.N. Zare, S. Sloss, and A.G. Ewing, Anal. Chem., 63 (1991) 189.
[7] X. Huang and R.N. Zare, Anal. Chem., 63 (1991) 2193.
[8] R.A. Wallingford and A.G. Ewing, Anal. Chem., 59 (1987) 1762.
[9] W. Lu and R.M. Cassidy, Anal. Chem., 66 (1994) 200.
[10] T.M. Olefirowicz and A.G. Ewing, Anal. Chem., 62 (1990) 1872.
[11] Th.J. O'Shea, R.D. Greenhagen, S.M. Lunte, C.E. Lunte, M.R. Smyth, D.M. Radzik and N. Watanabe, J. Chromatogr., 593 (1992) 305.
[12] I.C. Chen and C.-W. Whang, J. Chromatogr., 644 (1993) 208.
[13] W.Th. Kok and Y. Sahin, Anal. Chem., 65 (1993) 2497.
[14] X. Huang and W.Th. Kok, J. Chromatogr. A, 716 (1995) 347.
[15] W.Th. Kok, Anal. Chem., 65 (1993) 1853.
[16] F.D. Swanek, G.Y. Chen, and A.G. Ewing, Anal. Chem., 68 (1996) 3912.
[17] A.J. Tudos, M.M.C. van Dyck, H. Poppe and W.Th. Kok, Chromatographia, 37 (1993) 79.
[18] J. Ye and R.P. Baldwin, Anal. Chem., 65 (1993) 3525.
[19] D.K. Xu, L. Hua, and H.Y. Chen, Anal. Chim. Acta, 335 (1996) 95.
[20] A. Durgbanshi and W.Th. Kok, submitted to J. Chromatogr. A.
[21] B.L. Lin, L.A. Colon, and R.N. Zare, J. Chromatogr. A, 680 (1994) 263.
[22] M. Zhong, J. Zhou, S.M. Lunte, G. Zhao, D.M. Giolando and J.R. Kirchhoff, Anal. Chem., 68 (1996) 203.
[23] Th.J. O'Shea, S.M. Lunte, and W.R. LaCourse, Anal. Chem., 65 (1993) 948.
[24] W. Lu and R.M. Cassidy, Anal. Chem., 65 (1993) 2878.
[25] R.E. Roberts and D.C. Johnson, Electroanalysis, 7 (1995) 1015.
[26] P.L. Weber, T. Kornfelt, N.K. Klausen and S.M. Lunte, Anal. Biochem., 225 (1995) 135.
[27] W.R. Lacourse and G.S. Owens, Electrophoresis, 17 (1996) 310.
[28] L.A. Colon, R. Dadoo, and R.N. Zare, Anal. Chem., 65 (1993) 476.
[29] J.N. Ye and R.P. Baldwin, J. Chromatogr. A, 687 (1994) 141.
[30] Y. Guo, L.A. Colon, R. Dadoo, and R.N. Zare, Electrophoresis, 16 (1995) 493.
[31] J. Zhou and S.M. Lunte, Electrophoresis, 16 (1995) 498.
[32] A.M. Fermier and L.A. colon, HRC-J. High Resolut. Chromatogr., 19 (1996) 613.
[33] X.M. Fang, J.N. Ye, and Y.Z. Fang, Anal. Chim. Acta, 329 (1996) 49.
[34] T.J. O'Shea and S.M. Lunte, Anal. Chem., 65 (1993) 247.
[35] X. Huang and W.Th. Kok, J. Chromatogr. A, 707 (1995) 335.
[36] T.J. O'Shea and S.M. Lunte, Anal. Chem., 65 (1993) 247.
[37] F.E.P. Mikkers, F.M. Everaerts, and Th.P.E.M. Verheggen, J. Chromatogr., 169 (1979) 11.
[38] F. Foret, M. Deml, Y. Kahle, and P. Bocek, Electrophoresis, 7 (1993) 430.
[39] C. Haber, W.R. Jones, J. Soglia, M.A. Surve, M. Mcglynn, J.R. Reineck, and C. Krstanovic, J. Capillary Electrophoresis, 3 (1996) 1.
[40] N. Avdalovic, C.A. Pohl, R.D. Rocklin, and J.R. Stillian, Anal. Chem., 65 (1993) 1470.
[41] M. Harrold, J. Stillian, L. Bao, R. Rocklin, and N. Avdalovic, J. Chromatogr. A, 717 (1995) 371.

13 Micro-preparative CE

13.1 Fraction collection

Despite the miniaturised character of CE there is a strong interest in the preparative application of this separation technique [1,2]. The unsurpassed separation efficiency of CE, together with the fact that the separation takes place in free solution, makes this an interesting option. Of course bulk production is not the aim of such applications. Still, the production of small amounts of purified substances by CE is attractive. Separated fractions can be collected for sequencing purposes of DNA fragments or peptides [3,4], for off-line identification by mass spectrometry [5,6], to localise the biological activity of a preparation in a single substance, or for peak purity confirmation [7,8]. An example of this latter application is shown in Figure 13.1. A racemic mixture of terbutaline was separated by CE under overloaded conditions (Figure 13.1a). The reinjection of one of the collected fractions shows the enantiomeric purity of the product (Figure 13.1b).

With a number of commercially available instruments an option for fraction collection is present. This has been achieved by installing a programmable vial tray at the rear-end side of the capillary, similar to that installed in the autosampler at the front-end side. Shortly before an analyte zone of interest is expected to reach the end of the capillary, the voltage is switched off and the outlet buffer vial is replaced by a microvial. The vial should contain a small volume (in the order of 1 μL) of water or a dilute buffer solution to ensure that the high-voltage electrode can make contact. By reapplication of the voltage the zone is collected in this microvial. To increase the precision of timing, a lower voltage can be used during the fraction collection. After this, the vial is replaced again by a next microvial or by the outlet buffer vial. An accurate timing of the events is essential. In order to account for small changes in the migration times of zones (e.g., caused by variations of the electroosmotic mobility), on-column detection should be applied simultaneously, preferably at a position close to the end of the capillary.

A disadvantage of this so-called electro-elution method is that the voltage has to be switched on and off repeatedly during collection. This could affect the reproducibility of the migration times. An alternative collection device, with which the voltage can be kept on continuously, has been proposed by Mueller et al. [9] (see Figure 13.2). They developed a sheath flow connection for the end of the separation capillary. With fibre optics the BGE is monitored very close to the capillary end. The sheath liquid is merged with the ions migrating from the capillary by hydrodynamic pressure. Electrical contact with the BGE is made through the sheath solution, so that the voltage source can stay connected while the capillary system ends in "open air". The droplets containing the fractions of interest, formed at the PTFE tip of the system, are collected in 10 μL

glass capillaries. The droplets are sucked automatically into these glass capillaries by capillary action when brought in contact. The collection capillary's can be interchanged with the help of an automated rotor system. Figure 13.3 shows that with this system the 11 fragments of a *HAe*III restriction digest of ΦX-174 plasmid DNA can be collected separately; the reinjection of the fractions confirms that the cross-over is negligible.

13.2 Optimisation of the production rate

The mass capacity of a CE system is inherently limited. Under typical CE conditions, the maximum amount of analyte that can be purified in one run, without seriously compromising the separation efficiency, is in the order of 1 to 10 pmole [10]. Increasing the analyte concentration is limited by overloading effects, increasing the volume scale of the separation by excessive Joule heating, resulting in a general temperature elevation. A loss of separation efficiency by radial temperature gradients and by siphoning effects may also appear; in practice, however, these effects are usually not the bottleneck.

As has been discussed in Chapter 7, the presence of a relatively high concentration (compared to the BGE) of analyte ions in a zone disturbs the local field strength. Thereby, the apparent migration rate of the analyte in the zone (μ_{app}) becomes a function of its concentration. Triangular concentration profiles are developed that can be described by:

$$\mu_{app}(c) = \mu_0 \cdot \left(1 + \beta_{EMD} \cdot \frac{c}{c_b} \right) \quad (13.1)$$

where c is the local analyte concentration in the zone, c_b the salt concentration of the BGE, μ_0 the mobility of the analyte at infinite dilution, and β_{EMD} the so-called electromigration constant. Increasing the amount of analyte will lead to a wider range of the analyte concentration within in the zone, larger differences in apparent mobilities and therefore to a wider zone.

The (dimensionless) electromigration constant describes the susceptibility of an analyte/BGE combination for overloading. It depends on the acid/base and the electrophoretic properties of the analyte as well as the BGE constituents [11]. Electromigration dispersion can be minimised by a proper choice of the BGE composition. However, in practice it will generally not be possible to rule it out completely; under optimised conditions values of β_{EMD} in the order of 0.1 to 1 are realistic.

On basis of Equation (13.1) it can be derived that under overloading conditions the obtained plate number is inversely proportional to the amount of analyte injected. An approximate expression relating the maximum amount of analyte (n_{max}, in mole) that can be purified in one run to the plate number required for the separation (N_{req}) is [10]:

$$n_{max} \approx \frac{8}{N_{req}} \cdot \frac{1000c_b}{\beta_{EMD}} \cdot \frac{1}{4} \pi d_c^2 \cdot L \quad (13.2)$$

0009-5893/00   S-80-03 $ 03.00/0

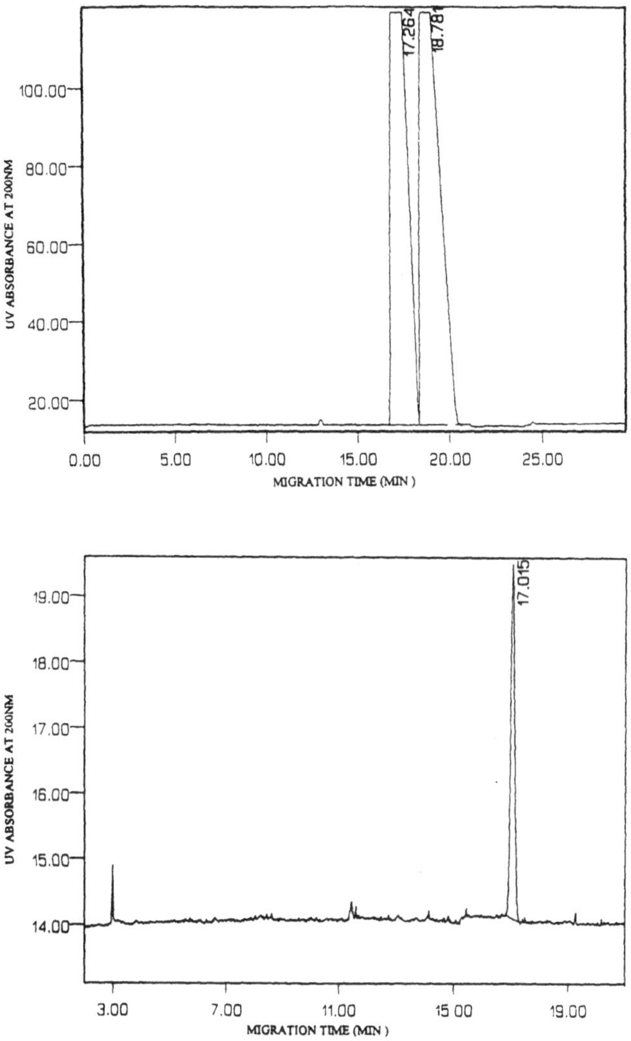

Figure 13.1

Preparative separation (a) and purity confirmation of a collected fraction (b) of a racemic terbutaline mixture. Reproduced from ref. [8].

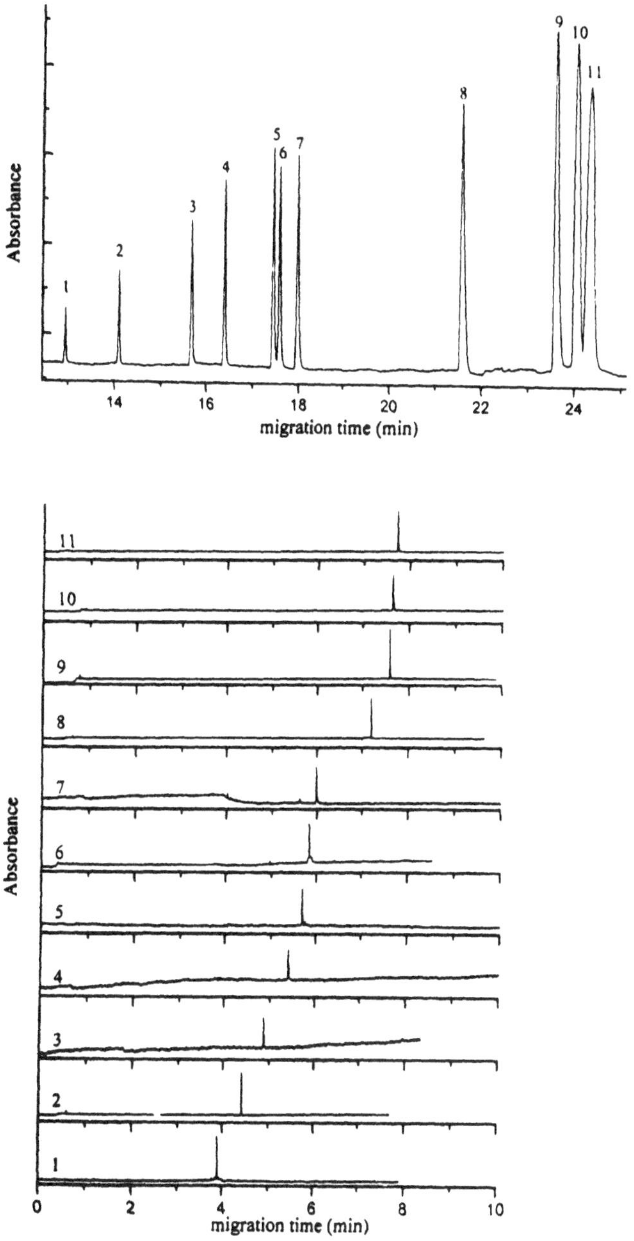

Figure 13.3

Separation (a) and reinjection of collected fractions (b) of ΦX-174/ HAeIII restriction fragments. Reproduced from ref. [9].

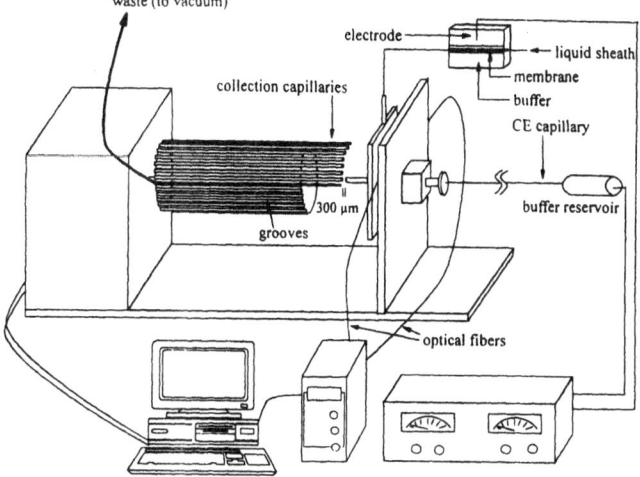

Figure 13.2

Schematic view of the sheath flow device for fraction collection. Reproduced from ref. [9].

Table 13.1 Comparison of the operating conditions and performances of analytical and a micropreparative CE systems.

application	analytical	preparative
V_{app} [kV]	25	10
L [m]	0.5	1
d_c [μm]	50	100
c_b [mol L^{-1}]	0.010	0.125
t_{run} [min]	6	60
N_{max}[a]	500,000	200,000
P [W m^{-1}]	0.5	1
ΔT [K]	5	10
n_{max}[b] [pmole]	2	200

[a] Plate number obtained without overloading.
[b] Maximum amount injected for N_{req}=100,000.

where the BGE salt concentration is in mol L^{-1}. Apparently, maximising the sample capacity of a CE system implies:

- optimising the BGE composition in respect to the susceptibility to overloading;
- increasing the BGE concentration;
- increasing the volume of the capillary.

When increasing the BGE concentration or the diameter of the capillary, however, other limitations will be met. It has been shown that the average temperature of the solution increases by Joule heating proportional to the electric power developed per unit length of the capillary (Equation 4.2). In preparative applications of CE, the problems to be expected from the temperature elevation are the decomposition of sample compounds or the disruption of the electrophoretic process by overheating of the solution. Therefore, the bottleneck is the part of the capillary with the least efficient cooling (the highest value of r). The power that can be applied is limited by the maximally allowed temperature in this part.

To maximise the mass capacity of a CE system, according to Equation 13.2 one would like to use a wide capillary and a high BGE salt concentration. However, when the power is bound by a maximum, this implies that the field strength applied should be limited. Apparently, the capacity of the separation system can be enlarged by increasing the length of the capillary, or decreasing the applied voltage. Of course, increasing L or decreasing V_{app} will result in longer run times. Still, even when the production rate (the amount of substance that can be purified per unit of time) is considered, the conclusion remains that for a high capacity the separation should be performed in long capillaries with a relatively low applied voltage. The run time that is regarded as still acceptable sets a limit here. A lower limit to V_{appl} is also set by the zone broadening due to axial diffusion; the

applied voltage should still be high enough to reach the required plate number (Equation 3.5).

In respect to the temperature increase as well as the mass capacity, it makes no difference whether a low BGE concentration is used in a wide capillary or a high concentration in a narrower capillary, as long as ($d_c^2 c_b$) is the same. In general, however, it is to be preferred to use a narrower capillary with a concentrated BGE in order to keep the possible siphoning of the solutions under control. The maximum solubility of the BGE and/or the sample compounds of interest are decisive here.

Summarising, in the optimisation of the micropreparative application of CE, one should:

- tune the composition of the BGE with the electrophoretic properties of the sample component to be purified, in order to obtain a low susceptibility to overloading (a low value of β_{EMD});
- use a high BGE concentration, within the limits set by the solubility of the buffer or the sample components;
- apply a relatively low voltage, but high enough to reach the required number of plates;
- use a long capillary, within the limit set by the acceptable run time;
- select a capillary with a wide diameter, within the limit set by the allowed temperature increase at a poorly thermostatted part of the capillary.

In Table 13.1 the operating conditions and performance of a typical analytical application of CE are compared to those of a (hypothetical) preparative application, optimised in the way presented above. Compared to the analytical system, the mass capacity per run of the preparative system is increased by a factor of 100; the production rate (in mole s^{-1}) is 10 times higher.

References

[1] D.J. Rose and J.W. Jorgenson, J. Chromatogr., 438 (1988) 23.
[2] A. Guttman, A.S. Cohen, D.N. Heiger and B.L. Karger, Anal. Chem., 62 (1990) 137.
[3] H.J. Boss, M.F. Rohde and R.S. Rush, Anal. Biochem., 230 (1995) 123.
[4] C. Schwer and F. Lottspeich, J. Chromatogr., 623 (1992) 345.
[5] N.H.H. Heegaard, and P. Roepstorff, J. Capillary Electrophor., 2 (1995) 219.
[6] H.G. Lee, J.L. Tseng, R.R. Becklin, and D.M. Desiderio, Anal. Biochem., 229 (1995) 188.
[7] K.D. Altria and Y.K. Dave, J. Chromatogr., 633 (1993) 221.
[8] K.D. Altria, Isolation and Purification, 2 (1996) 113.
[9] O. Mueller, F. Foret, and B.L. Karger, Anal. Chem., 67 (1995) 2974.
[10] A. Cifuentes, X. Xu, W.Th. Kok, and H. Poppe, J. Chromatogr. A, 716 (1995) 141.
[11] X. Xu, W.Th. Kok and H. Poppe, J. Chromatogr. A, 742 (1996) 211.

14 Instrumental Developments

14.1 Instruments for routine analysis

In the first years of the development of CE as a separation method, research groups had to build their own instruments. Since the basic equipment for CE can be quite simple, this was not a major problem. Even now, home-made instruments are used not only in academic groups but also occasionally in industrial laboratories [1]. For routine use, however, such devices are often not ideal, with an insufficient repeatability of the sample injection process as the main problem. Since 1986 a number of instrument manufacturers have stepped into the CE market. Apparently, the sales have not been up to expectation so far, and a few companies have already quit or have reduced their efforts and investments in CE. Some suppliers limit themselves to a specific application area. Waters for instance promotes its CE instrument as a Capillary Ion Analyzer (CIA), and the ABI PRISM® 310 and 3700 instruments are sold (by PE Biosystems) as Genetic Analyzers. Of course, there is nothing against using these instruments in other type of applications. Still, a number of suppliers offer general-purpose CE instruments; at present (end of 1999) in Europe these companies are Agilent (formerly known as the Hewlett Packard Chemical Analysis Group), Beckman Coulter, Bio-Rad Laboratories, Thermo BioAnalysis and Helena, which sells instruments produced by Prince Technologies.

Some instruments on the market are simple, modular R&D type apparatus, perfectly suited to learn the trade or to explore the possibilities of CE for a specific application. Most popular, however, certainly in routine laboratories, are complete, automated stand-alone systems. In the beginning of instrument development, manufacturers had to learn how to combine high-voltage electronics with sensitive microelectronics in a single instrument. This appeared to be non-trivial, and the reliability of the first commercial instruments was therefore not up to standard. While the customer satisfaction with the first-generation CE instruments may not have been overwhelming, presently it can be stated that the performance and reliability of most, if not all, instruments is satisfactory. The failure rate of the instruments appears to have been reduced to acceptable levels. Quantitative repeatability is comparable to that of the more mature technique HPLC, and software has been developed to deal with the variation of migration times inherent to CE. Recent innovation efforts have been focussed largely on interfacing and data processing software. Much attention has been given to the compliance of the instruments with modern regulatory issues. On the hardware side, progress has been limited in the previous years [2,3]. Most note-worthy is the improved possibility to couple a CE instrument to a mass spectrometer (see paragraph 14.2). However, very recently breakthroughs have been made in the fields of multiplex systems and in microfabricated CE systems. These developments will be discussed in the next paragraphs.

Table 14.1 Decision points for CE hardware and software

system part	features
high-voltage supply	• voltage upper limit • reversible polarity • constant voltage, current or power possibilities • recording of current and actual voltage during run
BGE supply	• free access to different BGE vials during run • BGE replenishment option • free access to different outlet vials
autosampler	• number of positions for sample vials • cooling and heating possibilities • compatible with micro titerplates
sample introduction	• hydrodynamic and electrokinetic injection • injection from multiple vials for on-column derivatization • high-precision pressure system with feedback system
capillary system	• free capillary or cassette system • effectiveness of thermostatting system (liquid or air) for micro-preparative applications • part of capillary length properly thermostatted • shortest length possible for high-speed applications
detector	• single or multiwavelength (diode-array or fast-scanning) absorbance detection • extended light path cell option • availability of other detection modes (LIF, conductivity) • possibility to use outside detector (MS) and minimum capillary length
pressure system	• high inlet pressure for rapid flushing and high-viscosity solutions (CGE) • high inlet and outlet pressure option for pressurized CEC • high-precision pressure system for zone mobilization in cIEF
data processing	• high acquisition rate possible • peak recognition and integration algorithms special for CE • simultaneous handling of positive and negative peaks • algorithm for identification of EOF peak • calculation of mobilities • peak identification on basis of mobilities • peak area correction with variable EOF velocity

Some apparatus on the market offer unique features that may be a reason to decide for them. For the Agilent ^{3D} Capillary Electrophoresis System extended light path capillaries and cells are available for high-sensitivity absorbance detection (see paragraph 9.4). The Beckman MDQ can accommodate industrial standard well plates, which allows the high-throughput analysis of up to 96 samples. Bio-Rad Laboratories offers a dual laser detector (the BioFocus® LIF^2 detector), with a 488 nm argon-ion and a 594 nm helium laser, and two independent data collection channels. The P/ACE– MDQ instrument from Beckman Coulter also features a dual wavelength laser-induced fluorescence option, but with an argon-ion laser only. The ABI Prism® 3700 instru-

0009-5893/00   S-83-07 $ 03.00/0

Table 14.2 CE instrument supplier websites

company	address
Agilent	http://www.chem.agilent.com/
Beckman-Coulter	http://www.beckmancoulter.com/
Bio-Rad Laboratories	http://www.bio-rad.com/
Helena	http://www.helena.com/
PE Biosystems	http://pebio.com/
Prince Technologies	http://ourworld.compuserve.com/home-pages/PrinceTechnologies/
Thermo BioAnalysis	http://www.biomolecular1.com/
Waters	http://www.waters.com/

ment can be equipped with a capillary array, enabling the simultaneous analysis of up to 96 samples (see paragraph 14.3). The Crystal 1000 detector (Thermo BioAnalysis) is the only conductivity detector commercially available at present (see paragraph 12.5). The Helena CES I system from Prince Technologies is in principle a modular system, very easy to couple to various detectors such as a lamp-based fluorometer (see paragraph 10.1) or a mass spectrometer (paragraph 14.2). The Waters instrument comes with a special mercury lamp for short-wavelength UV detection, providing high-sensitivity detection of proteins at 185 nm.

When special features as mentioned above are not required, there is an ample choice of suitable instruments. This is not the place to make a critical comparison of the instruments commercially available at present. In Table 14.1 a non-exhaustive list is given of hardware features that can be considered in the decision for a specific apparatus to be purchased.

In a routine environment the quality of the software for instrument control and data processing may be of equal importance as the quality of the hardware. Factors as user-friendliness of the interfacing, diagnostic aids or compliance with regulation regimes are important but not specific for a CE instrument. Some special software features to be considered when selecting a CE instrument are included in Table 14.1.

A novel trend in the CE instrumentation market is that suppliers do not offer just an instrument but an application. The instrument is then configured for this special purpose, and the offer may include capillaries, standards, reagents, and know-how. In such a case the quality of the service technicians and application chemists of the supplier will be an important factor when choosing between different offers.

A list of website addresses of current CE instrument suppliers (in Europe) is given in Table 14.2.

14.2 Coupling of CE and mass spectrometry

One of the major advances in instrumental development in CE over the last 10 years has been the introduction and commercialization of the coupling of CE with mass spectrometry (CE-MS). Progress in CE-MS has strongly profited from recent advances in MS sample in-troduction systems, especially from the introduction of the electrospray interface (ESI) by Fenn c.s. [4,5,6]. In the first successful attempt of CE-MS coupling (by Olivares et al. in 1987) an ESI was used [7]. Other types of on-line interfaces have also been studied, e.g., with continuous flow fast atom bombardment (CF-FAB) [8,9]. Matrix assisted laser desorption (MALDI) has mostly been used in an off-line mode so far [10], although successful attempts to couple MALDI on-line to a CE system, using a transport system, have recently been reported [11]. Still, the ESI appears to be the method of choice for most. Several CE instrument suppliers, sometimes in conjunction with an MS specialist, have brought CE-ESI-MS combinations on the market.

Several reviews on CE-MS have been published recently [12,13,14], and here only a short overview of the instrumental aspects of the interfacing will be given. When coupling a CE system to a mass spectrometer, a number of potential difficulties have to be overcome:

- The analyte molecules or ions have to be transferred from a liquid phase to, eventually, the high vacuum of the mass spectrometer, without the creation of an appreciable pressure difference over the separation capillary.

- The volume scale of the interface has to be compatible with the peak volumes encountered in high-efficiency CE.

- Some electrical contact has to be made to the "open end" of the capillary, to close the electrical circuit for the separation voltage.

- In some way the composition of the separation electrolyte has to be made compatible with the demands of the mass spectrometer.

The first two points raised above are accounted for when an electrospray is used as interface. In an ESI, droplet formation, solvent evaporation and ionization of the analyte molecules take place at atmospheric pressure. Only after ionisation the analytes are directed into the vacuum parts of the mass spectrometer. Nano-electrospray devices have been developed in first instance for low-volume MS analysis of, e.g., peptides in sequencing [15]. These are perfectly suited to accommodate the flow rates commonly encountered in CE (in the order of 100 nL/min). Certainly with the so-called pneumatically assisted sprays (in which a sheath gas stream is used to aid droplet formation) that are applied in most commercial instruments, the dead-volume between the spray point and the orifice of the MS is not a problem.

For the problem of making electrical contact with the capillary end several solutions have been proposed. Schematic representations of these different approaches are given in Figure 14.1, which has been reproduced from the review article by Banks [13]. Most often used is the sheath liquid interface that has been introduced by Smith c.s. [16]. Here, the separation capillary end is inserted into a coaxial metal tube. Between the separation capillary and the sheath tube the sheath liquid is pumped by means of a small syringe pump. Electrical contact with the separation buffer is made, through the sheath liquid, on the metal tube. Therefore, the sheath liquid has to have at least some electrical conductivity. However, since the distance to be bridged to the metal tube is only small, this conductivity may be low. Methanol or methanol/water mixtures can be used as sheet liquid, with a volatile acid (acetic acid) to improve the conductivity when necessary. The sheath liquid does not only provide a means to make electrical contact, it may also help the droplet formation, especially when the EOF flow rate from the separation capillary is low. The exact positioning of the separation capillary within the sheath flow tube and the sheath flow rate are of importance, both for the zone broadening as well as for the sensitivity of detection [17].

Sheathless interfaces have been proposed by a number of research groups [18,19,20]. In such an interface the tip of the capillary itself is the spray point. This tip should be sharpened or pulled to a fine point. On the outside of the capillary tip, a conductive coating (of gold or another metal) is deposited which serves as the electrical contact for the separation voltage source and for the spray voltage. A modification of this sheathless interface is the direct electrode interface (Figure 14.1 d). Electrical contact is made here with a thin gold wire directly inserted into the capillary tip [21].

A third approach is the liquid-junction interface (Figure 14.1 c), as introduced by Lee et al. [22. With this device the electric field used for the separation is decoupled at a point before the spray needle. The decoupling principle is similar to that used in conjunction with, e.g., amperometric detectors. An advantage of this approach is that the separation and spray voltages can be controlled completely independently. With the other set-ups, the separation and spray voltage sources always share a common lead.

Sheath liquid and sheathless interfaces both have certain advantages and disadvantages. With a sheath liquid, the solution properties can be altered after the separation to improve the spray properties. An organic modifier (methanol) is often added to decrease the surface tension; smaller droplets may be formed leading to an improved transfer efficiency of the interface. A volatile weak acid (formic, acetic) may be added to improve the ionisation process. On the other hand, with the sheath liquid the analyte zones will be diluted, and it has been reported that a sheathless interface gives higher sensitivities [23].The sheath liquid may also give problems with the separation. Ions from the sheath liquid solution or pH shifts may migrate into the separation capillary during the run, thereby altering the migration velocities of analytes and even the order of the peaks [24]. This will mainly be a problem with systems with a suppressed EOF such as used for protein separations. With a substantial EOF velocity in the direction of the interface, the entering of pH shifts into the capillary from the backside is easily avoided by choosing a proper buffer for the separation medium (see paragraph 4.3).

(a)
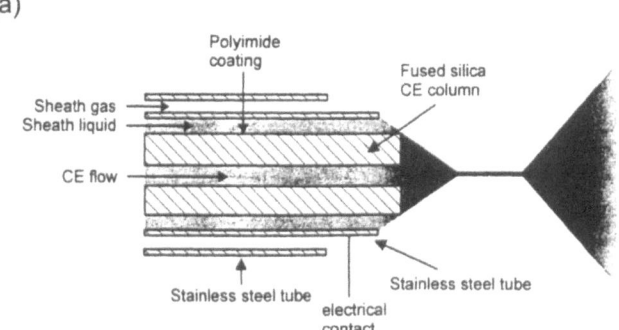

(b)
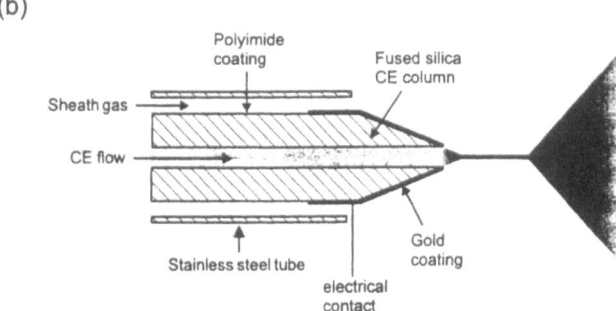

(c)
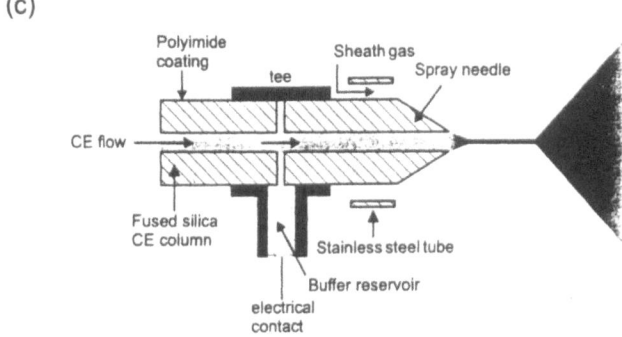

(d)
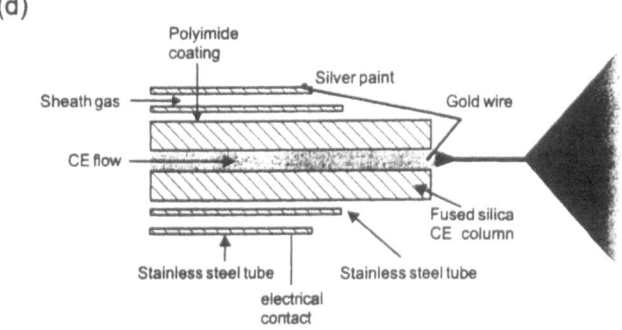

Figure 14.1
Different types of CE/ES interfaces: (A) sheath liquid, (B) sheathless, (C) liquid-junction; and (D) direct electrode. Reproduced form ref. [13].

In most applications of CE-ESI-MS so far, a quadrupole MS instrument has been used, probably because this type of instrument is the most widely available and best known to separation specialists. A quadrupole MS can even be used for the analysis of (bio)macromolecules such as proteins. The relatively low m/z limit of the quadrupole is compensated for by the ESI, which can produce multiple charged ions. However, the maximum scan rate of a quadrupole MS is in the order of 1 Hz. This

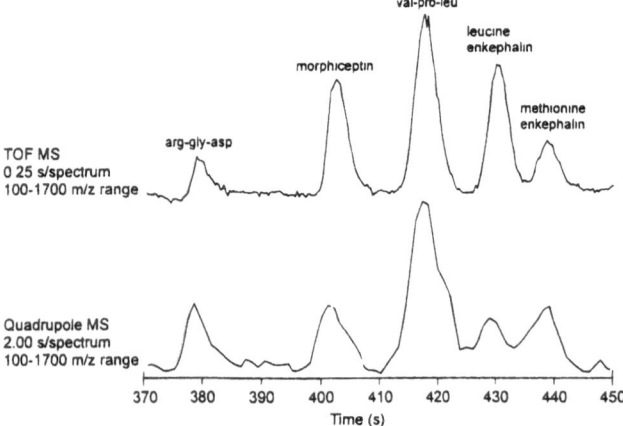

Figure 14.2
Comparison of the total ion currents from a peptide CE separation (bare silica capillary) using TOF and quadrupole instruments. Reproduced form ref. [17].

is actually too slow for a CE separation, where peaks are often only 1 or a few seconds wide. More adequate sampling rates can be realised with the "non-scanning" time-of flight (TOF) principle. Banks and Dresch [25] have shown that the resolution'in an electropherogram can be substantially improved when a TOF rather than a quadrupole MS is used (see Figure 14.2). For the coupling of a continuous ESI ion beam with the discrete sampling required when a TOF instrument is used, a number of solutions have been proposed. Of these, the orthogonal acceleration method [26] is probably the most popular [21,25]. Discrete samples from the ESI ion beam are drawn from it by applying electric field pulses perpendicular to the drift direction of the ionised analyte molecules. Pulsed ion packets then travel the flight tube to the detector. An alternative is to use an ion trap between the ESI and the TOF-MS [27]. In the ion trap, sample ions from the ESI beam are collected for a certain period of time and stored until they are pulsed as a packet into the mass spectrometer. Since the ion trap serves as a signal integrator high sensitivities can be obtained. A good example of this is shown in Figure 14.3. Digestion products of a protein could be detected by Wu et al. in the low fmole range with CE-ESI-ion trap-TOF-MS [28].

14.3 Capillary array CE

In the beginning of the development of CE, its high separation speed compared to conventional electrophoresis systems was acclaimed. Adherents of the slab gel methods argued, and quite rightly so, that they could simultaneously analyse a large number of samples, including standards and controls, on one apparatus. Therefore, the old methods could still be superior in terms of sample throughput.

Since one of the ideas behind CE was to have a fast method for high-throughput applications, such as for the Human Genome Project or genetic analysis in general [29], several research groups and instrument manufacturers have worked on the development of capillary array systems.

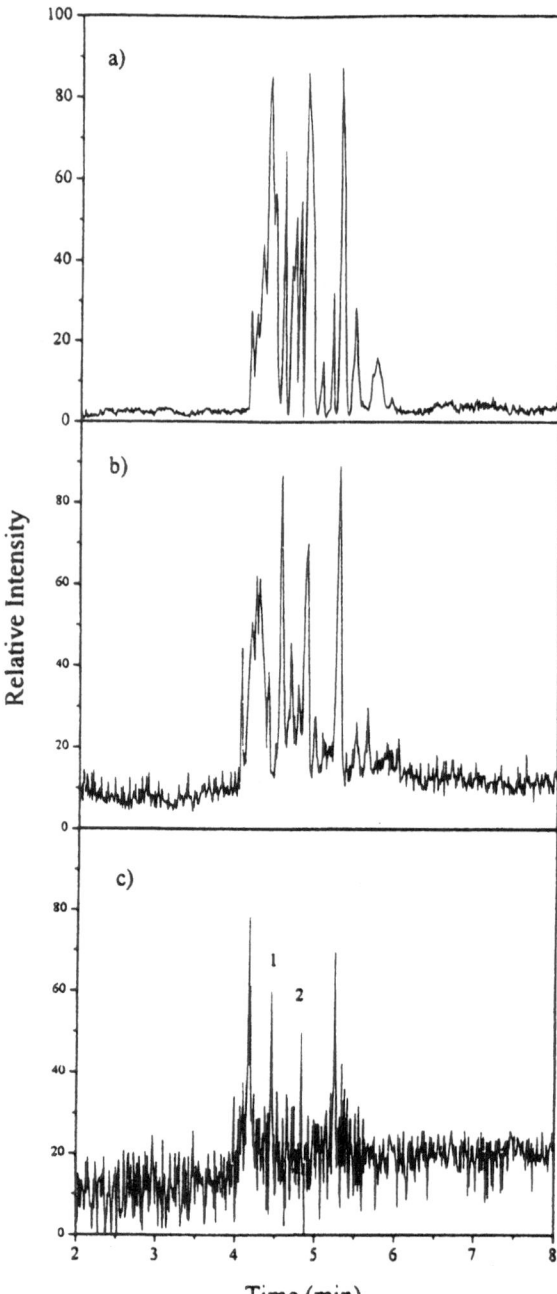

Figure 14.3
Total ion electropherograms of cytochrome *c* digest with sample injection amount of (a) 50, (b) 12, and (c) 3 fmole. Ion trap storage / reflectron TOF-MS; mass detection speed 4 Hz. Reproduced from ref. [28].

The premises for a capillary array system are of course that at least part of the instrumental set-up is shared: the voltage source, the sample introduction system and preferably part of the detection system. Sharing the detection system appeared to be the most problematic. Only very recently scientific research on multiplexed CE with UV absorbance detection has been published [30]. In almost all studies on capillary arrays published so far, laser-induced fluorescence has been applied for detection. This is apparently the easiest technique to be multiplexed. Different approaches have been studied to illuminate an array of capillaries with a single laser source:

Instrumental Developments

- The capillaries can be irradiated one at a time using a confocal fluorescence microscope as detector and an automated translation stage on which the array is fixed [31,32]. With such a laser scanning method the detection time has to be shared, which will have a negative effect on the signal-to-noise ratio.
- The laser beam can be split into different beams before hitting the separate capillaries. Taylor and Yeung used an optical fiber bundle for this [33]. On one end of the fiber bundle a laser was focused; the other end of the bundle was split and the single fibers were inserted axially into the ends of up to 10 capillaries in an array. Alternatively, the laser beam can be expanded in one direction and directed under a specific angle on a the parallel capillaries of the array [34]. By splitting of the laser beam, however, the power per capillary is decreased, which in some cases might lead to lower signal-to-noise ratios (see paragraph 10.2).
- By sideways illumination of the array, sharing of the laser power by the different capillaries will not be necessary. The optical system should be designed such that the different capillaries are illuminated with approximately the same intensity. This, however, is difficult because of the refraction and deflection of the light at the various inner and outer walls of the capillaries. In Figure 14.4 the laser-beam paths are shown calculated by computer simulation for various arrangements of the capillaries [35]. In the multiplexed LIF systems that are currently available on the market such sideways illumination is applied.

  Reflection and deflection problems are avoided when a multi-sheath-flow system is used as originally proposed by Kambara and Takahashi [36]; here, LIF detection is performed in the migration lanes in the solution beyond the ends of the capillaries. Such a system, however, is not too easy to implement in practice.

For detection in multiplexed LIF systems a charge-coupled device (CCD) is generally used. In one of the directions of such a two-dimensional detector array the different capillaries are imaged. The other direction can be used to obtain spectral information on the emitted light, for instance by using a set of optical filters. In Figure 14.5 a schematic view is given of the set-up to be used.

Presently two multiplexed CE systems are commercially available. In the ABI Prism® 3700 systems a 96-capillary array can be placed. With a multi-wavelength LIF detection system, this instrument is promoted for DNA analysis. For the Beckman MDQ an 8-capillary array is available. With a LIF detector it can be used for genetic analysis; with a low-wavelength UV lamp and absorbance detection it can be used for multiplexed serum protein analysis (the Paragon system).

14.4 Nanoscale integrated CE systems

Currently one of the areas of most active research is that of so-called CE-on-a-chip: the development of micromachined devices for CE separations. This work is part

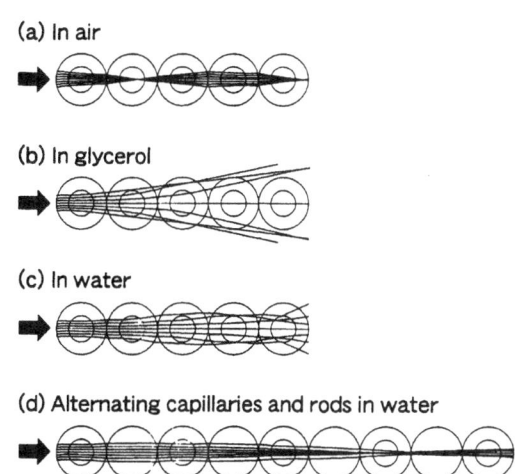

(a) In air

(b) In glycerol

(c) In water

(d) Alternating capillaries and rods in water

Figure 14.4
Computer-simulated laser-beam paths through different configurations of a capillary array. Reproduced from ref. [35].

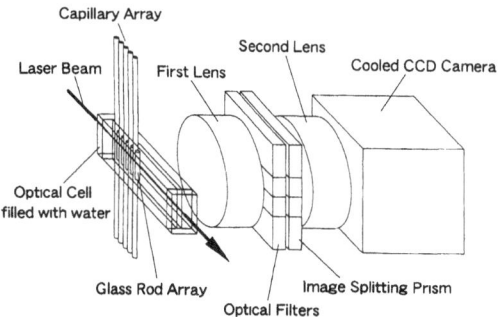

Figure 14.5
Schematic view of a capillary-array assembly and fluorescence detection system. Reproduced from ref. [35].

of the general pursuit of nanoscale, integrated analytical systems that can comply with the modern demands (low-cost, high throughput) of the chemical and pharmaceutical industry. It might have been more logical to try to transfer the work horse of separation science, HPLC, to nanoscale devices. However, suitable micro-pumps are not easily fabricated. On the other hand, liquid transport through electroosmosis, by applying an electrical field, is perfectly compatible with devices fabricated through micromachining. Microfabrication of CE systems using photolithographic technology was introduced by Manz c.s. in 1993 [37].

The normal substrate for chips in the electronic industry is silicon. This material, however, is not very suited for micromachined separation devices; its conductivity is too high, and its electrical breakdown limit too low, for the application of the usual CE voltages. Therefore, most studies focus on glass or quartz as basic material. Alternatively, polymer chips can be cast or molded using a silicon master [38,39]. Creating a separation channel in a chip wafer is not difficult. In fact, the dimensions of a normal CE channel ($10 - 50\ \mu$m) are fairly large for micromachining methods. Sample introduction can be performed in various ways; Figure 14.6 shows a way of fluid handling by electro-

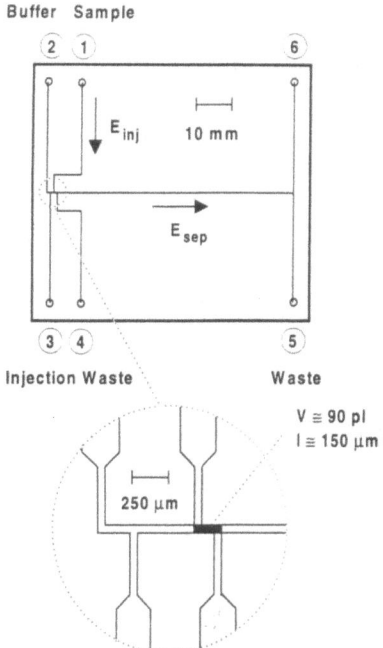

Figure 14.6
Channel system on an integrated CE glass chip for sample introduction. Adapted from ref. [41].

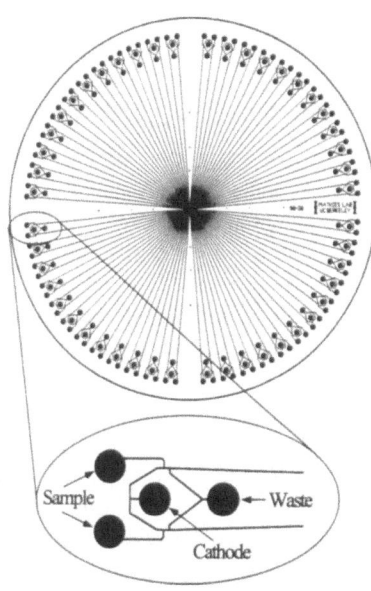

Figure 14.7
Pattern for the 96-channel radial capillary array electrophoresis microplate. The diameter of the reservoir holes is 1.2 mm; the distance from the injector to the detector point is 33 mm. Reproduced from ref. [40].

kinetic processes to accomplish the introduction of a minute volume of the sample solution. A recent design that allows the introduction and simultaneous separation of up to 96 samples is shown in Figure 14.7 [40].

Several reviews have appeared that report on the latest developments in this rapidly changing field [41,42]. Agilent has recently started marketing a 12-channel "lab-on-a-chip" device. It is intended for multiplexed DNA analysis, based on a size-dependent separation in an entangled polymer solution.

14.5 Future perspectives

The end of a monograph is a usual place for some tea-leaves diagnosis or crystal ball contemplation. Here, this tradition is respected.

The instrumentation available for middle-of-the-road CE applications appears to be mature and adequate. In terms of reproducibility, speed or (absorbance) detection sensitivity, no major breakthroughs are to be expected anymore. A wider choice of detection possibilities would come in handy. Lamp-based fluorescence detectors for instance could be wider available. There is also still a need for macromolecule-specific (light-scattering?) detection.

Of the more recent innovations in CE, ESI-MS will become more important and more the standard detection method in various application areas. While in LC-MS the quadrupole MS is the most often used so far, in CE-MS the TOF instruments may play a more important role. These instruments provide the acquisition speed and sensitivity that is required in CE.

Finally, there are the micromachined, integrated CE systems. The human resources and investments put in this field are huge, and important advances have been made. On the other hand the pH ISFET, that was supposed to replace the old-fashioned glass pH electrode, took over 20 years to develop from an R&D device to a real product, and even then it was not much of a success. Apparently, it is easier to juggle with electrons on chips than it is with ions or solutions. One of the points that has to be addressed urgently is the detection with this nanoscale technology. It may be nice to be able to do a (multiplex) separation on a 5×5 cm platform, when for detection a present-day deep-UV laser, which is not known for its portability, has to be used part of the fun is spoilt. Nanoscale, integrated analytical devices, on which electroosmotic and electrophoretic processes will play an important role, undoubtedly have a future in the real world. However, when this future will start: the author of this review has no idea.

References

[1] K.D. Altria and S.M. Bryant, LC-GC International, 10 (1997) 26.
[2] M. Warner, Anal. Chem. 66 (1994) 1137A.
[3] C. Henry, Anal. Chem., 68 (1996) 747A.
[4] M. Yamashita and J.B. Fenn, J. Phys. Chem., 88 (1984) 4451.
[5] C.M. Whitehouse, R.N. Dryer, M. Yamashita and J.B. Fenn, Anal. Chem., 57 (1985) 675.
[6] J.B. Fenn, M. Mann, C.K. Meng, S.F. Wong and C.M. Whitehouse, Science, 246 (1989) 64.
[7] J.A. Olivares, N.T. Nguyen, C.R. Yonker and R.D. Smith, Anal. Chem., 59 (1989) 1230.
[8] R.D. Minard, D. Luckenbill, P.J. Curry and A.G. Ewing, Adv. Mass Spectrom., 11 (1989) 436.
[9] M.A. Moseley, L.J. Deterding, K.B. Tomer and J.W. Jorgenson, J. Chromatogr., 516 (1989) 197.
[10] F. Foret, O. Mueller, J. Thorne, W. Goetzinger and B.L. Karger, J. Chromatogr. A, 716 (1995) 157.
[11] J. Preisler, F. Foret and B.L. Karger, Anal. Chem., 70 (1998) 5278.

[12] J. Cai and J. Henion, J. Chromatogr. A, 703 (1995) 667.
[13] J.F. Banks, Electrophoresis, 18 (1997) 2255.
[14] J. Ding and P. Vouros, Anal. Chem., 71 (1999) 378A.
[15] M.S. Wilms and M. Mann, Int. J. Mass Spectrom. Ion Processes, 136 (1994) 167.
[16] R.D. Smith, C.J. Barinaga, and H.R. Udseth, Anal. Chem., 60 (1988) 1948.
[17] J.F. Banks and T. Dresch, Anal. Chem., 68 (1996) 1480.
[18] J.H. Wahl, D.C. Gale and R.D. Smith, J. Chromatogr., 659 (1994) 217.
[19] M.S. Kriger, K.D. Cook, and R.S. Ramsey, Anal. Chem., 67 (1995) 385.
[20] G.A. Valaskovic, N.L. Kelleher, and F.W. McLafferty, Science, 273 (1996) 1270.
[21] L. Fang, R. Zhang, E.R. Williams, and R.N. Zare, Anal. Chem., 66 (1994) 3696.
[22] E.D. Lee, W. Muck, J.D. Henion, and T.R. Covey, J. Chromatogr., 458 (1988) 313.
[23] D.C. Gale and R.D. Smith, Rapid Commun. Mass Spectrom., 7 (1993) 1017.
[24] F. Foret, T.J. Thompson, P. Vouros, B.L. Karger, P. Gebauer, and P. Bocek, Anal. Chem., 66 (1994) 4450.
[25] J.F. Banks and Th. Dresch, Anal. Chem., 68 (1996) 1480.
[26] J.H.J. Dawson and M. Guilhaus, Rapid Commun. Mass Spectrom., 3 (1989) 155.
[27] J.-T. Wu, M.G. Qian, M.X. Li, K. Zheng, P. Huang, and D.M. Lubman, J. Chromatogr. A, 794 (1998) 377.
[28] J.-T. Wu, M.G. Qian, M.X. Li, L. Liu, and D.M. Lubman, Anal. Chem., 68 (1996) 3388.
[29] I. Kheterpal and R.A. Mathies, Anal. Chem., 71 (1999) 31A.
[30] X. Gong and E.S. Yeung, Anal. Chem., 71 (1999) 4989.
[31] R.A. Mathies and X.C. Huang, Nature, 359 (1992) 167.
[32] J.R. Scherer, I. Kheterpal, A. Radhakrishnan, W.W. Ja and R.A. Mathies, Electrophoresis, 20 (1999) 1508.
[33] J.A. Taylor and E.A. Yeung, Anal. Chem., 65 (1993) 956.
[34] K. Ueno and E.A. Yeung, Anal. Chem., 66 (1994) 1424.
[35] T. Anazawa, S. Takahashi and H. Kambara, Electrophoresis, 20 (1999) 539.
[36] H. Kambara and S. Takahashi, Nature, 361 (1993) 565.
[37] D.J. Harrison, K. Fluri, K. Seiler, Z. Fan, C.S. Effenhauser, and A. Manz, Science, 261 (1993) 895.
[38] C.S. Effenhauser, G.J.M. Bruin, A. Paulus, and M. Ehrat, Anal. Chem., 69 (1997) 3451.
[39] R.M. McCornick, R.J. Nelson, M.G. Alonso-Amigo, D.J. Benvegnu, and H.H. Hooper, Anal. Chem., 69 (1997) 2626.
[40] Y. Shi, P.C. Simpson, J.R. Scherer, D. Wexler, C. Skibola, M.T. Smith, and R.A. Mathies, Anal. Chem., 71 (1991) 5354.
[41] C.S. Effenberger, G.J.M. Bruin, and A. Paulus, Electrophoresis, 18 (1997) 2203.
[42] F.E. Regnier, Chromatographia, Supplement I, 49 (1999) S-56.